20**24** ISPSC®

INTERNATIONAL **SWIMMING POOL AND SPA** CODE®

INTERNATIONAL CODE COUNCIL

POOL & HOT TUB ALLIANCE

ICC
INTERNATIONAL CODE COUNCIL®

2024 International Swimming Pool and Spa Code®

First Printing: July 2023

ISBN: 978-1-959851-91-2 (soft-cover edition)
ISBN: 978-1-959851-92-9 (PDF download)

COPYRIGHT © 2023
by
INTERNATIONAL CODE COUNCIL, INC.

T029748

PRINTED IN THE USA

NEW DESIGN FOR THE 2024 INTERNATIONAL CODES

IBC IRC IFC IPC IMC IECC IEBC IFGC IPMC IPSDC IWUIC IZC ICCPC IgCC ISPSC

The 2024 International Codes® (I-Codes®) have undergone substantial formatting changes as part of the digital transformation strategy of the International Code Council® (ICC®) to improve the user experience. The resulting product better aligns the print and PDF versions of the I-Codes with the ICC's Digital Codes® content.

The changes, promoting a cleaner, more modern look and enhancing readability and sustainability, include:

Streamlined lists

QR codes to identify code changes more accurately
(For further details, see Formatting Changes to the 2024 International Codes.)

Consistent grouping of associated content
(e.g., tables immediately follow parent sections)

Single-column text

Shading for table headers and notes

A B C D
a b c d
Modernized font styles

More information can be found at iccsafe.org/design-updates.

PREFACE

FORMATTING CHANGES TO THE 2024 INTERNATIONAL CODES

The 2024 International Codes® (I-Codes®) have undergone substantial formatting changes as part of the digital transformation strategy of the International Code Council® (ICC®) to improve the user experience. The resulting product better aligns the print and PDF versions of the I-Codes with the ICC's Digital Code content. Additional information can be found at iccsafe.org/design-updates.

Replacement of Marginal Markings with QR Codes

Through 2021, print editions of the I-Codes identified technical changes from prior code cycles with marginal markings [solid vertical lines for new text, deletion arrows (➡), asterisks for relocations (⁕)]. The 2024 I-Code print editions replace the marginal markings with QR codes to identify code changes more precisely.

A QR code is placed at the beginning of any section that has undergone technical revision. If there is no QR code, there are no technical changes to that section.

In the following example from the 2024 *International Swimming Pool and Spa Code*® (ISPSC®), a QR code indicates there are changes to Section 101 from the 2021 ISPSC. Note that the change may occur in the main section or in one or more subsections of the main section.

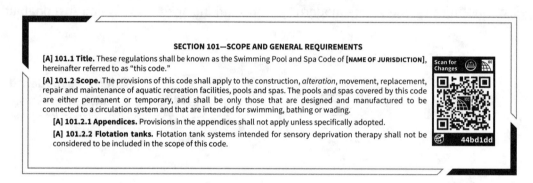

To see the code changes, the user need only scan the QR code with a smart device. If scanning a QR code is not an option, changes can be accessed by entering the 7-digit code beneath the QR code at the end of the following URL: qr.iccsafe.org/ (in the above example, "qr.iccsafe.org/44bd1dd"). Those viewing the code book via PDF can click on the QR code.

All methods take the user to the appropriate section on ICC's Digital Codes website, where technical changes from the prior cycle can be viewed. Digital Codes Premium subscribers who are logged in will be automatically directed to the Premium view. All other users will be directed to the Digital Codes Basic free view. Both views show new code language in blue text along with deletion arrows for deleted text and relocation markers for relocated text.

Digital Codes Premium offers additional ways to enhance code compliance research, including revision histories, commentary by code experts and an advanced search function. A full list of features can be found at codes.iccsafe.org/premium-features.

ACCESSING ADDITIONAL FEATURES VIA REGISTRATION OF BOOK

Beginning with the 2024 *International Mechanical Code*® (IMC®) and the 2024 *International Plumbing Code*® (IPC®), users will be able to validate the authenticity of their book and register it with the ICC to receive incentives. Digital Codes Premium (codes.iccsafe.org) provides advanced features and exclusive content to enhance code compliance. To validate and register, the user will tap the ICC tag (pictured here and located on the front cover) with a near-field communication (NFC) compatible device. Visit iccsafe.org/nfc for more information and troubleshooting tips regarding NFC tag technology.

ABOUT THE I-CODES

The 2024 I-Codes, published by the ICC, are 15 fully compatible titles intended to establish provisions that adequately protect public health, safety and welfare; that do not unnecessarily increase construction costs; that do not restrict the use of new materials, products or methods of construction; and that do not give preferential treatment to particular types or classes of materials, products or methods of construction.

The I-Codes are updated on a 3-year cycle to allow for new construction methods and technologies to be incorporated into the codes. Alternative materials, designs and methods not specifically addressed in the I-Code can be approved by the building official where the proposed materials, designs or methods comply with the intent of the provisions of the code.

The I-Codes are used as the basis of laws and regulations in communities across the US and in other countries. They are also used in a variety of nonregulatory settings, including:

- Voluntary compliance programs.
- The insurance industry.

- Certification and credentialing for building design, construction and safety professionals.
- Certification of building and construction-related products.
- Facilities management.
- "Best practices" benchmarks for designers and builders.
- College, university and professional school textbooks and curricula.
- Reference works related to building design and construction.

Code Development Process

The code development process regularly provides an international forum for building professionals to discuss requirements for building design, construction methods, safety, performance, technological advances and new products. Proposed changes to the I-Codes, submitted by code enforcement officials, industry representatives, design professionals and other interested parties, are deliberated through an open code development process in which all interested and affected parties may participate.

Openness, transparency, balance, due process and consensus are the guiding principles of both the ICC Code Development Process and OMB Circular A-119, which governs the federal government's use of private-sector standards. The ICC process is open to anyone without cost. Remote participation is available through cdpAccess®, the ICC's cloud-based app.

In order to ensure that organizations with a direct and material interest in the codes have a voice in the process, the ICC has developed partnerships with key industry segments that support the ICC's important public safety mission. Some code development committee members were nominated by the following industry partners and approved by the ICC Board:

- American Gas Association (AGA)
- American Institute of Architects (AIA)
- American Society of Plumbing Engineers (ASPE)
- International Association of Fire Chiefs (IAFC)
- National Association of Home Builders (NAHB)
- National Association of State Fire Marshals (NASFM)
- National Council of Structural Engineers Association (NCSEA)
- National Multifamily Housing Council (NMHC)
- Plumbing Heating and Cooling Contractors (PHCC)
- Pool and Hot Tub Alliance (PHTA) formerly The Association of Pool and Spa Professionals (APSP)

Code development committees evaluate and make recommendations regarding proposed changes to the codes. Their recommendations are then subject to public comment and council-wide votes. The ICC's governmental members—public safety officials who have no financial or business interest in the outcome—cast the final votes on proposed changes.

The I-Codes are subject to change through future code development cycles and by any governmental entity that enacts the code into law. For more information regarding the code development process, contact the Codes and Standards Development Department of the ICC at iccsafe.org/products-and-services/i-codes/code-development/.

While the I-Code development procedure is thorough and comprehensive, the ICC, its members and those participating in the development of the codes expressly disclaim any liability resulting from the publication or use of the I-Codes, or from compliance or noncompliance with their provisions. NO WARRANTY OF ANY KIND, IMPLIED, EXPRESSED OR STATUTORY, IS GIVEN WITH RESPECT TO THE I-CODES. The ICC does not have the power or authority to police or enforce compliance with the contents of the I-Codes.

Code Development Committee Responsibilities (Letter Designations in Front of Section Numbers)

In each cycle, proposed changes are considered by the Code Development Committee assigned to a specific code or subject matter. Committee Action Hearings result in recommendations regarding a proposal to the voting membership. Where changes to a code section are not considered by that code's own committee, the code section is preceded by a bracketed letter designation identifying a different committee. Bracketed letter designations for the I-Code committees are:

[A] = Administrative Code Development Committee
[BE] = IBC—Egress Code Development Committee
[BF] = IBC—Fire Safety Code Development Committee
[BG] = IBC—General Code Development Committee
[BS] = IBC—Structural Code Development Committee
[E] = Developed under the ICC's Standard Development Process
[EB] = International Existing Building Code Development Committee
[F] = International Fire Code Development Committee
[FG] = International Fuel Gas Code Development Committee
[M] = International Mechanical Code Development Committee

[P]   =   International Plumbing Code Development Committee

[SP]   =   International Swimming Pool and Spa Code Development Committee

For the development of the 2027 edition of the I-Codes, the ICC Board of Directors approved a standing motion from the Board Committee on the Long-Term Code Development Process to revise the code development cycle to incorporate two committee action hearings for each code group. This change expands the current process from two independent 1-year cycles to a single continuous 3-year cycle. There will be two groups of code development committees and they will meet in separate years. The current groups will be reworked. With the energy provisions of the *International Energy Conservation Code*® (IECC®) and Chapter 11 of the *International Residential Code*® (IRC®) now moved to the Code Council's Standards Development Process, the reduced volume of code changes will be distributed between Groups A and B.

Code change proposals submitted for code sections that have a letter designation in front of them will be heard by the respective committee responsible for such code sections. Because different committees hold Committee Action Hearings in different years, proposals for most codes will be heard by committees in both the 2024 (Group A) and the 2025 (Group B) code development cycles. It is very important that anyone submitting code change proposals understands which code development committee is responsible for the section of the code that is the subject of the code change proposal.

Please visit the ICC website at iccsafe.org/products-and-services/i-codes/code-development/current-code-development-cycle for further information on the Code Development Committee responsibilities as it becomes available.

Coordination of the I-Codes

The coordination of technical provisions allows the I-Codes to be used as a complete set of complementary documents. Individual codes can also be used in subsets or as stand-alone documents. Some technical provisions that are relevant to more than one subject area are duplicated in multiple model codes.

Italicized Terms

Words and terms defined in Chapter 2, Definitions, are italicized where they appear in code text and the Chapter 2 definitions apply. Although care has been taken to ensure applicable terms are italicized, there may be instances where a defined term has not been italicized or where a term is italicized but the definition found in Chapter 2 is not applicable. For example, Chapter 2 of the *International Building Code*® (IBC®) contains a definition for "*Listed*" that is applicable to equipment, products and services. The term "listed" is also used in that code to refer to a list of items within the code or within a referenced document. For the latter, the Chapter 2 definition would not be applicable.

Adoption of International Code Council Codes and Standards

The International Code Council maintains a copyright in all of its codes and standards. Maintaining copyright allows the Code Council to fund its mission through sales of books in both print and digital format. The Code Council welcomes incorporation by reference of its codes and standards by jurisdictions that recognize and acknowledge the Code Council's copyright in the codes and standards, and further acknowledge the substantial shared value of the public/private partnership for code development between jurisdictions and the Code Council. By making its codes and standards available for incorporation by reference, the Code Council does not waive its copyright in its codes and standards.

The Code Council's codes and standards may only be adopted by incorporation by reference in an ordinance passed by the governing body of the jurisdiction. "Incorporation by reference" means that in the adopting ordinance, the governing body cites only the title, edition, relevant sections or subsections (where applicable), and publishing information of the model code or standard, and the actual text of the model code or standard is not included in the ordinance (see graphic, "Adoption of International Code Council Codes and Standards"). The Code Council does not consent to the reproduction of the text of its codes or standards in any ordinance. If the governing body enacts any changes, only the text of those changes or amendments may be included in the ordinance.

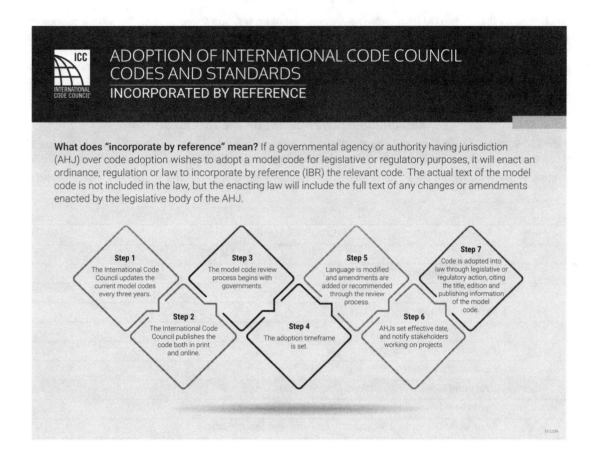

ADOPTION OF INTERNATIONAL CODE COUNCIL CODES AND STANDARDS
INCORPORATED BY REFERENCE

What does "incorporate by reference" mean? If a governmental agency or authority having jurisdiction (AHJ) over code adoption wishes to adopt a model code for legislative or regulatory purposes, it will enact an ordinance, regulation or law to incorporate by reference (IBR) the relevant code. The actual text of the model code is not included in the law, but the enacting law will include the full text of any changes or amendments enacted by the legislative body of the AHJ.

Step 1
The International Code Council updates the current model codes every three years.

Step 2
The International Code Council publishes the code both in print and online.

Step 3
The model code review process begins with governments.

Step 4
The adoption timeframe is set.

Step 5
Language is modified and amendments are added or recommended through the review process.

Step 6
AHJs set effective date, and notify stakeholders working on projects.

Step 7
Code is adopted into law through legislative or regulatory action, citing the title, edition and publishing information of the model code.

The Code Council also recognizes the need for jurisdictions to make laws accessible to the public. Accordingly, all I-Codes and I-Standards, along with the laws of many jurisdictions, are available to view for free at codes.iccsafe.org/codes/i-codes. These documents may also be purchased, in both digital and print versions, at shop.iccsafe.org.

To facilitate adoption, some I-Code sections contain blanks for fill-in information that needs to be supplied by the adopting jurisdiction as part of the adoption legislation. For example, the ISPSC contains:

Section 101.1. Insert: [NAME OF JURISDICTION]

Section 103.1. Insert: [NAME OF DEPARTMENT]

Section 113.4. Insert: [SPECIFY OFFENSE], [AMOUNT], [NUMBER OF DAYS]

Section A101.3. Insert: [NUMBER OF YEARS]

For further information or assistance with adoption, including a sample ordinance, jurisdictions should contact the Code Council at incorporation@iccsafe.org.

For a list of frequently asked questions (FAQs) addressing a range of foundational topics about the adoption of model codes by jurisdictions and to learn more about the Code Council's code adoption resources, scan the QR code or visit iccsafe.org/code-adoption-resources.

INTRODUCTION TO THE INTERNATIONAL SWIMMING POOL AND SPA CODE

The ISPSC establishes minimum requirements for the design, construction, alteration, repair and maintenance of swimming pools, spas, hot tubs and aquatic facilities. This includes public swimming pools, public spas, public exercise spas, aquatic recreation facilities, onground storable residential pools, permanent inground residential pools, permanent residential spas, permanent residential exercise spas, portable residential spas and portable residential exercise spas.

In many jurisdictions, in addition to code officials having the responsibility for reviewing plans and inspecting the construction of pools and spas, environmental health officials also have a responsibility for oversight of the operation of pools and spas. In order to prevent disease and injuries, environmental health officials conduct operational evaluations (inspections). This may include water chemistry, credentials and training of pool operators and lifeguards, proper water circulation, facility staff's preparedness to respond to injuries and accidents, and proper sanitation and safety of the facility.

Code officials and environmental health officials commonly work closely in the plan review and inspection of pools and spas. This collaboration between departments to jointly review plans and inspect pools and spas is critical in order to achieve a safe and healthy environment for all that utilize these facilities.

The Pool and Hot Tub Alliance (PHTA) [formerly The Association of Pool & Spa Professionals (APSP)], a cooperating sponsor with ICC in the development and update of the ISPSC, further notes: "While it is recognized that proper construction and installation are essential, safe use of pools and spas requires common sense, including constant adult supervision of children and proper operation and maintenance. It is assumed and intended that pool users will exercise appropriate personal judgment and responsibility and that owners and operators of public and residential pools and spas will create and enforce rules and warning appropriate for their facilities."

ARRANGEMENT AND FORMAT OF THE 2024 ISPSC

The format of the ISPSC allows each chapter to be devoted to a particular subject with the exception of Chapter 3, which contains general compliance subject matter that is coordinated with the provisions for each type of pool and spa regulated in Chapters 4–10.

The following table shows how the ISPSC is divided. The chapter synopses detail the scope and intent of the provisions of the ISPSC.

CHAPTER TOPICS	
CHAPTERS	**SUBJECTS**
1	Scope and Administration
2	Definitions
3	General Compliance
4	Public Swimming Pools
5	Public Spas and Public Exercise Spas
6	Aquatic Recreation Facilities
7	Onground Storable Residential Swimming Pools
8	Permanent Inground Residential Swimming Pools
9	Permanent Residential Spas and Permanent Residential Exercise Spas
10	Portable Residential Spas and Portable Residential Exercise Spas
11	Referenced Standards
Appendix A	Board of Appeals

Chapter 1 Scope and Administration.

Chapter 1 establishes the limits of applicability of the code and describes how the code is to be applied and enforced. The provisions of Chapter 1 establish the authority and duties of the code official appointed by the authority having jurisdiction and also establish the rights and privileges of the design professional, contractor and property owner.

Chapter 2 Definitions.

Chapter 2 is the repository of the definitions of terms used in the body of the code. The user of the code should be familiar with and consult this chapter because the definitions are essential to the correct interpretation of the code and because the user may not be aware that a term is defined.

Chapter 3 General Compliance.

Chapter 3 includes a variety of requirements for pools and spas intended to maintain a minimum level of safety and sanitation for both the general public and the users of pools or spas.

Chapter 4 Public Swimming Pools.

Chapter 4 sets forth specific requirements for public swimming pools with regard to diving equipment, bather load limitations, rest ledges, wading pools, decks, deck equipment, filters, dressing and sanitary facilities, special features and signage.

Chapter 5 Public Spas and Public Exercise Spas.

Chapter 5 establishes the specific criteria for public spas and public exercise spas with regard to materials, structure and design, pumps and motors, return and suction fittings, heater and temperature requirements, water supply, sanitation, oxidation equipment and chemical feeders, and safety features.

Chapter 6 Aquatic Recreation Facilities.

Chapter 6 establishes specific requirements for aquatic recreation facilities with regard to floors, markings and indications, circulation systems, handholds and ropes, depths, barriers, number of occupants, toilet rooms and bathrooms, special features and signage.

Chapter 7 Onground Storable Residential Swimming Pools.

Chapter 7 establishes specific requirements for onground storable residential swimming pools with regard to ladders and stairs, decks and circulation systems.

Chapter 8 Permanent Inground Residential Swimming Pools.

Chapter 8 establishes specific requirements for permanent inground residential swimming pools with regard to design, construction tolerances, diving water envelopes, walls, offset ledges, pool floors, diving equipment, special features, circulation systems and safety features.

Chapter 9 Permanent Residential Spas and Permanent Residential Exercise Spas.

Chapter 9 establishes specific requirements for permanent residential spas and permanent residential exercise spas with regard to safety features.

Chapter 10 Portable Residential Spas and Portable Residential Exercise Spas.

Chapter 10 establishes specific requirements for portable residential spas and portable residential exercise spas with regard to standards that the equipment must meet.

Chapter 11 Referenced Standards.

Chapter 11 lists all of the product and installation standards and codes that are referenced throughout Chapters 1 through 10 and includes identification of the promulgators and the section numbers in which the standards and codes are referenced. As stated in Section 102.7, these standards and codes become an enforceable part of the code (to the prescribed extent of the reference) as if printed in the body of the code.

Appendix A Board of Appeals.

Appendix A contains the provisions for appeal and the establishment of a board of appeals. The provisions include the application for an appeal, the makeup of the board of appeals and the conduct of the appeal process.

RELOCATION OF TEXT OR TABLES

The following table indicates relocation of sections and tables in the 2024 edition of the ISPSC from the 2021 edition.

RELOCATIONS	
2024 LOCATION	2021 LOCATION
104.2.2.4	104.11
104.2.3	104.10
104.2.4	104.9
104.4	104.6
104.7.2	104.5
104.7.4	104.11.3
107	106
108	107
109	108
110	109
111	110
112	111

CONTENTS

SCOPE AND ADMINISTRATION

PART 1—SCOPE AND APPLICATION

SECTION 101—SCOPE AND GENERAL REQUIREMENTS

[A] 101.1 Title. These regulations shall be known as the Swimming Pool and Spa Code of [NAME OF JURISDICTION], hereinafter referred to as "this code."

[A] 101.2 Scope. The provisions of this code shall apply to the construction, *alteration*, movement, replacement, repair and maintenance of aquatic recreation facilities, pools and spas. The pools and spas covered by this code are either permanent or temporary, and shall be only those that are designed and manufactured to be connected to a circulation system and that are intended for swimming, bathing or wading.

    [A] 101.2.1 Appendices. Provisions in the appendices shall not apply unless specifically adopted.

    [A] 101.2.2 Flotation tanks. Flotation tank systems intended for sensory deprivation therapy shall not be considered to be included in the scope of this code.

[A] 101.3 Purpose. The purpose of this code is to establish minimum requirements to provide a reasonable level of safety, health, property protection and general welfare by regulating and controlling the design, construction, installation, quality of materials, location and maintenance or use of pools and spas.

[A] 101.4 Severability. If any section, subsection, sentence, clause or phrase of this code is for any reason held to be unconstitutional, such decision shall not affect the validity of the remaining portions of this code.

SECTION 102—APPLICABILITY

[A] 102.1 General. Where there is a conflict between a general requirement and a specific requirement, the specific requirement shall govern. Where, in any specific case, different sections of this code specify different materials, methods of construction or other requirements, the most restrictive shall govern.

[A] 102.2 Existing installations. Any pool or spa and related mechanical, electrical and plumbing systems lawfully in existence at the time of the adoption of this code shall be permitted to have their use and maintenance continued if the use, maintenance or repair is in accordance with the original design and no hazard to life, health or property is created.

[A] 102.3 Maintenance. Pools and spas and related mechanical, electrical and plumbing systems, both existing and new, and parts thereof, shall be maintained in proper operating condition in accordance with the original design in a safe and sanitary condition. Devices or safeguards that are required by this code shall be maintained in compliance with the edition of the code under which they were installed.

    The owner or the owner's authorized agent shall be responsible for maintenance of systems. To determine compliance with this provision, the *code official* shall have the authority to require any system to be reinspected.

[A] 102.4 Alterations or repairs. *Alterations* or *repairs* to any pool, spa or related system shall conform to that required for a new system without requiring the existing systems to comply with the requirements of this code. Alterations or repairs shall not cause existing systems to become unsafe, insanitary or overloaded.

    Minor alterations and repairs to existing systems shall be permitted in the same manner and arrangement as in the existing system, provided that such repairs or replacement are not hazardous and are *approved*.

[A] 102.5 Historic buildings. The provisions of this code relating to the construction, *alteration*, repair, enlargement, restoration, relocation or moving of pools, spas or systems shall not be mandatory for existing pools, spas or systems identified and classified by the state or local jurisdiction as part of a historic structure where such pools, spas or systems are judged by the *code official* to be safe and in the public interest of health, safety and welfare regarding any proposed construction, *alteration*, repair, enlargement, restoration, relocation or moving of such pool or spa.

[A] 102.6 Moved pools and spas. Except as determined by Section 102.2, systems that are a part of a pool, spa or system moved into or within the jurisdiction shall comply with the provisions of this code for new installations.

[A] 102.7 Referenced codes and standards. The codes and standards referenced in this code shall be those that are listed in Chapter 11 and such codes and standards shall be considered to be part of the requirements of this code to the prescribed extent of each such reference. Where differences occur between provisions of this code and the referenced standards, the provisions of this code shall be the minimum requirements.

[A] 102.7.1 Application of the International Codes. Where the *International Residential Code* is referenced in this code, the provisions of the *International Residential Code* shall apply to related systems in detached one- and two-family dwellings and townhouses not more than three stories in height. Other related systems shall comply with the applicable International Code or referenced standard.

[A] 102.8 Requirements not covered by code. Any requirements necessary for the strength, stability or proper operation of an existing or proposed system, or for the public safety, health and general welfare, not specifically covered by this code shall be determined by the *code official*.

[A] 102.9 Other laws. The provisions of this code shall not be deemed to nullify any provisions of local, state or federal law.

[A] 102.10 Application of references. References to chapter or section numbers, or to provisions not specifically identified by number, shall be construed to refer to such chapter, section or provision of this code.

PART 2—ADMINISTRATION AND ENFORCEMENT

SECTION 103—CODE COMPLIANCE AGENCY

[A] 103.1 Creation of agency. The [NAME OF DEPARTMENT] is hereby created and the official in charge thereof shall be known as the *code official*. The function of the agency shall be the implementation, administration and enforcement of the provisions of this code.

[A] 103.2 Appointment. The *code official* shall be appointed by the chief appointing authority of the jurisdiction.

[A] 103.3 Deputies. In accordance with the prescribed procedures of this jurisdiction and with the concurrence of the appointing authority, the *code official* shall have the authority to appoint a deputy *code official*, other related technical officers, inspectors and other employees. Such employees shall have powers as delegated by the *code official*.

SECTION 104—DUTIES AND POWERS OF THE CODE OFFICIAL

[A] 104.1 General. The *code official* is hereby authorized and directed to enforce the provisions of this code.

[A] 104.2 Determination of compliance. The code official shall have the authority to determine compliance with this code, to render interpretations of this code and to adopt policies and procedures in order to clarify the application of its provisions. Such interpretations, policies and procedures:

1. Shall be in compliance with the intent and purpose of this code.
2. Shall not have the effect of waiving requirements specifically provided for in this code.

[A] 104.2.1 Listed compliance. Where this code or a referenced standard requires equipment, materials, products or services to be listed and a listing standard is specified, the listing shall be based on the specified standard. Where a listing standard is not specified, the listing shall be based on an approved listing criteria. Listings shall be germane to the provision requiring the listing. Installation shall be in accordance with the listing and the manufacturer's instructions, and where required to verify compliance, the listing standard and manufacturer's instructions shall be made available to the code official.

[A] 104.2.2 Technical assistance. To determine compliance with this code, the code official is authorized to require the owner or owner's authorized agent to provide a technical opinion and report.

[A] 104.2.2.1 Cost. A technical opinion and report shall be provided without charge to the jurisdiction.

[A] 104.2.2.2 Preparer qualifications. The technical opinion and report shall be prepared by a qualified engineer, specialist, laboratory or specialty organization acceptable to the code official. The code official is authorized to require design submittals to be prepared by, and bear the stamp of, a registered design professional.

[A] 104.2.2.3 Content. The technical opinion and report shall analyze the properties of the design, operation or use of the building or premises and the facilities and appurtenances situated thereon to identify and propose necessary recommendations.

[A] 104.2.2.4 Tests. Where there is insufficient evidence of compliance with the provisions of this code, the code official shall have the authority to require tests as evidence of compliance. Test methods shall be as specified in this code or by other recognized test standards. In the absence of recognized test standards, the code official shall approve the testing procedures. Such tests shall be performed by a party acceptable to the code official.

[A] 104.2.3 Alternative materials, design and methods of construction and equipment. The provisions of this code are not intended to prevent the installation of any material or to prohibit any design or method of construction not specifically prescribed by this code, provided that any such alternative is not specifically prohibited by this code and has been approved.

Exception: Performance-based alternative materials, designs or methods of construction and equipment complying with the *International Code Council Performance Code*.

[A] 104.2.3.1 Approval authority. An alternative material, design or method of construction shall be approved where the code official finds that the proposed alternative is satisfactory and complies with Sections 104.2.3 through 104.2.3.7, as applicable.

[A] 104.2.3.2 Application and disposition. Where required, a request to use an alternative material, design or method of construction shall be submitted in writing to the code official for approval. Where the alternative material, design or method of construction is not approved, the code official shall respond in writing, stating the reasons the alternative was not approved.

[A] 104.2.3.3 Compliance with code intent. An alternative material, design or method of construction shall comply with the intent of the provisions of this code.

[A] 104.2.3.4 Equivalency criteria. An alternative material, design or method of construction shall, for the purpose intended, be not less than the equivalent of that prescribed in this code with respect to all of the following, as applicable:

1. Quality.
2. Strength.
3. Effectiveness.
4. Durability.
5. Safety, other than fire safety.
6. Fire safety.

[A] 104.2.3.5 Tests. Tests conducted to demonstrate equivalency in support of an alternative material, design or method of construction application shall be of a scale that is sufficient to predict performance of the end use configuration. Tests shall be performed by a party acceptable to the code official.

[A] 104.2.3.6 Reports. Supporting documentation, where necessary to assist in the approval of materials or assemblies not specifically provided for in this code, shall comply with Sections 104.2.3.6.1 and 104.2.3.6.2.

[A] 104.2.3.6.1 Evaluation reports. Evaluation reports shall be issued by an approved agency and use of the evaluation report shall require approval by the code official for the installation. The alternate material, design or method of construction and product evaluated shall be within the scope of the code official's recognition of the approved agency. Criteria used for the evaluation shall be identified within the report and, where required, provided to the code official.

[A] 104.2.3.6.2 Other reports. Reports not complying with Section 104.2.3.6.1 shall describe criteria, including but not limited to any referenced testing or analysis, used to determine compliance with code intent and justify code equivalence. The report shall be prepared by a qualified engineer, specialist, laboratory or specialty organization acceptable to the code official. The code official is authorized to require design submittals to be prepared by, and bear the stamp of, a registered design professional.

[A] 104.2.3.7 Peer review. The code official is authorized to require submittal of a peer review report in conjunction with a request to use an alternative material, design or method of construction, prepared by a peer reviewer that is approved by the code official.

[A] 104.2.4 Modifications. Where there are practical difficulties involved in carrying out the provisions of this code, the code official shall have the authority to grant modifications for individual cases provided that the code official shall first find that one or more special individual reasons make the strict letter of this code impractical, that the modification is in compliance with the intent and purpose of this code, and that such modification does not lessen health, accessibility, life and fire safety or structural requirements. The details of the written request for and action granting modifications shall be recorded and entered in the files of the department of building safety.

[A] 104.2.4.1 Flood hazard areas. The code official shall not grant modifications to any provision required in flood hazard areas as established by Section 1612.3 of the *International Building Code* unless a determination has been made that:

1. A showing of good and sufficient cause that the unique characteristics of the size, configuration or topography of the site render the elevation standards of Section 1612 of the *International Building Code* inappropriate.
2. A determination that failure to grant the variance would result in exceptional hardship by rendering the lot undevelopable.
3. A determination that the granting of a variance will not result in increased flood heights, additional threats to public safety or extraordinary public expense; cause fraud on or victimization of the public; or conflict with existing laws or ordinances.
4. A determination that the variance is the minimum necessary to afford relief, considering the flood hazard.
5. Submission to the applicant of written notice specifying the difference between the design flood elevation and the elevation to which the building is to be built, stating that the cost of flood insurance will be commensurate with the increased risk resulting from the reduced floor elevation, and stating that construction below the design flood elevation increases risks to life and property.

[A] 104.3 Applications and permits. The code official shall receive applications, review construction documents and issue permits, inspect the premises for which such permits have been issued and enforce compliance with the provisions of this code.

[A] 104.3.1 Determination of substantially improved or substantially damaged existing buildings and structures in flood hazard areas. For applications for reconstruction, rehabilitation, repair, alteration, addition or other improvement of existing buildings or structures located in flood hazard areas, the code official shall determine if the proposed work constitutes substantial improvement or repair of substantial damage. Where the code official determines that the proposed work constitutes substantial improvement or repair of substantial damage, and where required by this code, the code official shall require the building to meet the requirements of Section 1612 of the *International Building Code* or Section R322 of the *International Residential Code*, as applicable.

[A] 104.4 Right of entry. Where it is necessary to make an inspection to enforce the provisions of this code, or where the *code official* has reasonable cause to believe that there exists in a structure or on a premises a condition that is contrary to or in violation of this code that makes the structure or premises unsafe, dangerous or hazardous, the *code official* is authorized to enter the structure or premises at reasonable times to inspect or to perform the duties imposed by this code. If such structure or premises is occupied, the code official shall present credentials to the occupant and request entry. If such structure or premises is unoccupied, the code official shall first make a reasonable effort to locate the owner, the owner's authorized agent or other person having charge or control of the structure or premises and request entry. If entry is refused, the code official shall have recourse to every remedy provided by law to secure entry.

[A] 104.4.1 Warrant. Where the code official has first obtained a proper inspection warrant or other remedy provided by law to secure entry, an owner, the owner's authorized agent or occupant or person having charge, care or control of the structure or premises shall not fail or neglect, after a proper request is made as herein provided, to permit entry therein by the code official for the purposes of inspection and examination pursuant to this code.

[A] 104.5 Identification. The *code official* shall carry proper identification when inspecting structures or premises in the performance of duties under this code.

[A] 104.6 Notices and orders. The code official shall issue necessary notices or orders to ensure compliance with this code. Notices of violations shall be in accordance with Section 113.

[A] 104.7 Official records. The code official shall keep official records as required by Sections 104.7.1 through 104.7.5. Such official records shall be retained for not less than 5 years or for as long as the building or structure to which such records relate remains in existence, unless otherwise provided by other regulations.

[A] 104.7.1 Approvals. A record of approvals shall be maintained by the code official and shall be available for public inspection during business hours in accordance with applicable laws.

[A] 104.7.2 Inspections. The code official shall have the authority to conduct inspections, or shall accept reports of inspection by approved agencies or individuals. Reports of such inspections shall be in writing and be certified by a responsible officer of such approved agency or by the responsible individual. The code official shall keep a record of each inspection made, including notices and orders issued, showing the findings and disposition of each.

[A] 104.7.3 Code alternatives and modifications. Application for alternative materials, design and methods of construction and equipment in accordance with Section 104.2.3; modifications in accordance with Section 104.2.4; and documentation of the final decision of the code official for either shall be in writing and shall be retained in the official records.

[A] 104.7.4 Tests. The code official shall keep a record of tests conducted to comply with Sections 104.2.2.4 and 104.2.3.5.

[A] 104.7.5 Fees. The code official shall keep a record of fees collected and refunded in accordance with Section 109.

[A] 104.8 Liability. The code official, member of the board of appeals or employee charged with the enforcement of this code, while acting for the jurisdiction in good faith and without malice in the discharge of the duties required by this code or other pertinent law or ordinance, shall not thereby be personally liable, either civilly or criminally, and is hereby relieved from personal liability for any damage accruing to persons or property as a result of any act or by reason of any act or omission in the discharge of official duties.

[A] 104.8.1 Legal defenses. Any suit or criminal complaint instituted against any officer or employee because of an act performed by that officer or employee in the lawful discharge of duties under the provisions of this code or other laws or ordinances implemented through the enforcement of this code shall be defended by legal representatives of the jurisdiction until the final termination of the proceedings. The code official or any subordinate shall not be liable for costs in any action, suit or proceeding that is instituted in pursuance of the provisions of this code.

[A] 104.9 Approved materials and equipment. Materials, equipment and devices approved by the code official shall be constructed and installed in accordance with such approval.

[A] 104.9.1 Materials and equipment reuse. Materials, equipment and devices shall not be reused unless such elements are in good working condition and approved.

SECTION 105—PERMITS

[A] 105.1 Where required. Any owner, or owner's authorized agent who desires to construct, enlarge, alter, repair, move, or demolish a pool or spa or to erect, install, enlarge, alter, repair, remove, convert or replace any system, the installation of which is regulated by this code, or to cause any such work to be performed, shall first make application to the *code official* and obtain the required permit for the work. A permit shall not be required for replastering or resurfacing of an existing pool or spa.

[A] 105.2 Application for permit. Each application for a permit, with the required fee, shall be filed with the *code official* on a form furnished for that purpose and shall contain a general description of the proposed work and its location. The application shall be signed by the owner or the owner's authorized agent. The permit application shall contain such other information required by the *code official*.

[A] 105.3 Time limitation of application. An application for a permit for any proposed work shall be deemed to have been abandoned 180 days after the date of filing unless such application has been pursued in good faith or a permit has been issued; except that the *code official* is authorized to grant one or more extensions of time for additional periods not exceeding 180 days each. The extension shall be requested in writing and justifiable cause demonstrated.

[A] 105.4 Permit issuance. The application, construction documents and other data filed by an applicant for permit shall be reviewed by the *code official*. If the *code official* finds that the proposed work conforms to the requirements of this code and laws and ordinances applicable thereto, and that the fees specified in Section 109.1 have been paid, a permit shall be issued to the applicant.

[A] 105.4.1 Approved construction documents. When the *code official* issues the permit where construction documents are required, the construction documents shall be endorsed in writing and stamped "APPROVED." Such *approved* construction documents shall not be changed, modified or altered without authorization from the *code official*. Work shall be done in accordance with the *approved* construction documents.

The *code official* shall have the authority to issue a permit for the construction of a part of a system before the entire construction documents for the whole system have been submitted or *approved*, provided that adequate information and detailed statements have been filed complying with pertinent requirements of this code. The holders of such permit shall proceed at their own risk without assurance that the permit for the entire system will be granted.

[A] 105.4.2 Validity. The issuance of a permit or approval of construction documents shall not be construed to be a permit for, or an approval of, any violation of any of the provisions of this code or any other ordinance of the jurisdiction. Any permit presuming to give authority to violate or cancel the provisions of this code shall not be valid.

The issuance of a permit based on construction documents and other data shall not prevent the *code official* from thereafter requiring the correction of errors in said construction documents and other data or from preventing building operations being carried on thereunder where in violation of this code or of other ordinances of this jurisdiction.

[A] 105.4.3 Expiration. Every permit issued shall become invalid unless the work authorized by such permit is commenced within 180 days after its issuance, or if the work authorized by such permit is suspended or abandoned for a period of 180 days after the time the work is commenced. The *code official* is authorized to grant, in writing, one or more extensions of time, for a period not more than 180 days. The extension shall be requested in writing and justifiable cause demonstrated.

[A] 105.4.4 Extensions. Any permittee holding an unexpired permit shall have the right to apply for an extension of the time within which the permittee will commence work under that permit when work is unable to be commenced within the time required by this section for good and satisfactory reasons. The *code official* shall extend the time for action by the permittee for a period not exceeding 180 days if there is reasonable cause. The fee for an extension shall be one-half the amount required for a new permit for such work.

[A] 105.4.5 Suspension or revocation of permit. The *code official* shall revoke a permit or approval issued under the provisions of this code in case of any false statement or misrepresentation of fact in the application or on the construction documents on which the permit or approval was based.

SECTION 106—TEMPORARY STRUCTURES, EQUIPMENT AND SYSTEMS

[A] 106.1 General. The code official is authorized to issue a permit for temporary structures, equipment or systems. Such permits shall be limited as to time of service, but shall not be permitted for more than 180 days. The code official is authorized to grant extensions for demonstrated cause.

[A] 106.2 Conformance. Temporary structures, equipment and systems shall conform to the requirements of this code as necessary to ensure health, safety and general welfare.

[A] 106.3 Temporary service utilities. The code official is authorized to give permission to temporarily supply service utilities in accordance with Section 110.

[A] 106.4 Termination of approval. The code official is authorized to terminate such permit for a temporary structures, equipment, or system and to order the same to be discontinued.

SECTION 107—CONSTRUCTION DOCUMENTS

[A] 107.1 Construction documents. Construction documents, engineering calculations, diagrams and other such data shall be submitted in two or more sets with each application for a permit. The *code official* shall require construction documents, computations and specifications to be prepared and designed by a registered design professional where required by state law. Construction documents shall be drawn to scale and shall be of sufficient clarity to indicate the location, nature and extent of the work proposed and show in detail that the work conforms to the provisions of this code.

[A] 107.2 Retention of construction documents. One set of *approved* construction documents shall be retained by the *code official* for a period of not less than 180 days from date of completion of the permitted work, or as required by state or local laws. One set of *approved* construction documents shall be returned to the applicant, and said set shall be kept on the site of the building or work at all times during which the work authorized thereby is in progress.

SECTION 108—NOTICE OF APPROVAL

[A] 108.1 Approval. After the prescribed tests and inspections indicate that the work complies in all respects with this code, a notice of approval shall be issued by the *code official*.

[A] 108.2 Revocation. The *code official* is authorized to, in writing, suspend or revoke a notice of approval issued under the provisions of this code wherever the notice is issued in error, or on the basis of the incorrect information supplied, or where it is determined that the building or structure, premise, system or portion thereof is in violation of any ordinance or regulation or any of the provisions of this code.

SECTION 109—FEES

[A] 109.1 Fees. A permit shall not be valid until the fees prescribed by law have been paid. An amendment to a permit shall not be released until the additional fee, if any, has been paid.

[A] 109.2 Schedule of permit fees. Where work requires a permit, a fee for each permit shall be paid as required, in accordance with the schedule as established by the applicable governing authority

[A] 109.3 Permit valuations. The applicant for a permit shall provide an estimated value of the work for which the permit is being issued at time of application. Such estimated valuations shall include total value of work, including materials and labor, for which the permit is being issued, such as mechanical equipment and permanent systems. Where, in the opinion of the *code official*, the valuation is underestimated, the permit shall be denied unless the applicant can show detailed estimates to the *code official*. The code official shall have the authority to adjust the final valuation for permit fees.

Scan for Changes

97395cf

[A] 109.4 Work commencing before permit issuance. Any person who commences any work on a mechanical system before obtaining the necessary permits shall be subject to a fee as established by the *code official* that shall be in addition to the required permit fees.

[A] 109.5 Related fees. The payment of the fee for the construction, *alteration*, removal or demolition for work done in connection to or concurrently with the work authorized by a permit shall not relieve the applicant or holder of the permit from the payment of other fees that are prescribed by law.

[A] 109.6 Refunds The *code official* is authorized to establish a refund policy.

SECTION 110—SERVICE UTILITIES

[A] 110.1 Connection of service utilities. A person shall not make connections from a utility, source of energy, fuel, power, water system or sewer system to any building or system that is regulated by this code for which a permit is required until authorized by the *code official*.

[A] 110.2 Temporary connection. The *code official* shall have the authority to authorize the temporary connection of the building or system to the utility, source of energy, fuel, power, water system or sewer system for the purpose of testing systems or for use under a temporary approval.

[A] 110.3 Authority to disconnect service utilities. The *code official* shall have the authority to authorize disconnection of utility service to the building, structure or system regulated by this code and the referenced codes and standards in case of emergency where necessary to eliminate an immediate hazard to life or property or where such utility connection has been made without the approval required by Section 110.1 or 110.2. The *code official* shall notify the serving utility, and wherever possible the owner or the owner's authorized agent and occupant of the building, structure or service system of the decision to disconnect prior to taking such action. If not notified prior to disconnecting, the owner, the owner's authorized agent or occupant of the building, structure or service system shall be notified in writing, as soon as practical thereafter.

SECTION 111—INSPECTIONS

[A] 111.1 General. Construction or work for which a permit is required shall be subject to inspection by the *code official* and such construction or work shall remain visible and able to be accessed for inspection purposes until *approved*. Approval as a result of an inspection shall not be construed to be an approval of a violation of the provisions of this code or of other ordinances of the jurisdiction. Inspections presuming to give authority to violate or cancel the provisions of this code or of other ordinances of the jurisdiction shall not be valid. It shall be the duty of the permit applicant to cause the work to remain available and exposed for inspection purposes. Neither the *code official* nor the jurisdiction shall be liable for expense entailed in the removal or replacement of any material required to allow inspection.

Scan for Changes

0a03b8d

[A] 111.2 Preliminary inspection. Before issuing a permit, the *code official* is authorized to examine or cause to be examined buildings, structures and sites for which an application has been filed.

[A] 111.3 Required inspections and testing. Pool and spa installations or *alterations* thereto, including equipment, piping, and appliances related thereto, shall be inspected by the *code official* to ensure compliance with the requirements of this code.

[A] 111.4 Other inspections. In addition to the inspections specified in Sections 111.2 and 111.3, the *code official* is authorized to make or require other inspections of any construction work to ascertain compliance with the provisions of this code and other laws that are enforced.

[A] 111.5 Inspection request. It shall be the duty of the holder of the permit or their duly authorized agent to notify the *code official* when work is ready for inspection. It shall be the duty of the permit holder to provide access to and means for inspections of such work that are required by this code.

[A] 111.6 Approval required. Work shall not be done beyond the point indicated in each successive inspection without first obtaining the approval of the *code official*. The *code official,* upon notification, shall make the requested inspection and shall either indicate the portion of the construction that is satisfactory as completed, or notify the permit holder or his or her agent wherein the same fails to comply with this code. Any portions that do not comply shall be corrected and such portion shall not be covered or concealed until authorized by the *code official*.

[A] 111.7 Approved agencies. Test reports submitted to the *code official* for consideration shall be developed by *approved* agencies that have satisfied the requirements as to qualifications and reliability.

[A] 111.8 Evaluation and follow-up inspection services. Prior to the approval of a closed, prefabricated system and the issuance of a permit, the *code official* shall require the submittal of an evaluation report on each prefabricated system indicating the complete details of the system, including a description of the system and its components, the basis on which the system is being evaluated, test results and similar information, and other data as necessary for the *code official* to determine conformance to this code.

[A] 111.9 Evaluation service. The *code official* shall designate the evaluation service of an *approved* agency as the evaluation agency, and review such agency's evaluation report for adequacy and conformance to this code.

[A] 111.10 Follow-up inspection. Except where ready access is provided to systems, service equipment and accessories for complete inspection at the site without disassembly or dismantling, the *code official* shall conduct the frequency of in-plant inspections necessary to ensure conformance to the *approved* evaluation report or shall designate an independent, *approved* inspection agency to conduct such inspections. The inspection agency shall furnish the *code official* with the follow-up inspection manual and a report of inspections on request, and the system shall have an identifying label permanently affixed to the system indicating that factory inspections have been performed.

[A] 111.11 Test and inspection records. Required test and inspection records shall be available to the *code official* at all times during the fabrication of the system and the installation of the system, or such records as the *code official* designates shall be filed.

[A] 111.12 Special inspections. Special inspections of alternative engineered design systems shall be conducted in accordance with Sections 111.12.1 and 111.12.2.

> **[A] 111.12.1 Periodic inspection.** The registered design professional or designated inspector shall periodically inspect and observe the alternative engineered design to determine that the installation is in accordance with the *approved* construction documents. Discrepancies shall be brought to the immediate attention of the contractor for correction. Records shall be kept of inspections.

> **[A] 111.12.2 Written report.** The registered design professional shall submit a final report in writing to the *code official* upon completion of the installation, certifying that the alternative engineered design conforms to the *approved* construction documents. A notice of approval for the system shall not be issued until a written certification has been submitted.

[A] 111.13 Testing. Systems shall be tested as required by this code. Tests shall be made by the permit holder and the *code official* shall have the authority to witness such tests.

[A] 111.14 New, altered, extended or repaired systems. New systems and parts of existing systems that have been altered, extended or repaired shall be tested as prescribed by this code.

[A] 111.15 Equipment, material and labor for tests. Equipment, material and labor required for testing a system or part thereof shall be furnished by the permit holder.

[A] 111.16 Reinspection and testing. Where any work or installation does not pass any initial test or inspection, the necessary corrections shall be made to comply with this code. The work or installation shall then be resubmitted to the *code official* for inspection and testing.

SECTION 112—MEANS OF APPEALS

Scan for Changes

c7caaaf

[A] 112.1 General. In order to hear and decide appeals of orders, decisions or determinations made by the *code official* relative to the application and interpretation of this code, there shall be and is hereby created a board of appeals. The board of appeals shall be appointed by the applicable governing authority and shall hold office at its pleasure. The board shall adopt rules of procedure for conducting its business and shall render all decisions and findings in writing to the appellant with a duplicate copy to the *code official*.

[A] 112.2 Limitations on authority. An application for appeal shall be based on a claim that the true intent of this code or the rules legally adopted thereunder have been incorrectly interpreted, the provisions of this code do not fully apply, or an equivalent or better form of construction is proposed. The board shall not have authority to waive requirements of this code.

[A] 112.3 Qualifications. The board of appeals shall consist of members who are qualified by experience and training on matters pertaining to the provisions of this code and are not employees of the jurisdiction.

[A] 112.4 Administration. The *code official* shall take action without delay in accordance with the decision of the board.

SECTION 113—VIOLATIONS

[A] 113.1 Unlawful acts. It shall be unlawful for any person, firm or corporation to erect, construct, alter, repair, remove, demolish or utilize any system, or cause same to be done, in conflict with or in violation of any of the provisions of this code.

[A] 113.2 Notice of violation. The *code official* shall serve a notice of violation or order to the person responsible for the erection, installation, *alteration*, extension, repair, removal or demolition of work in violation of the provisions of this code, or in violation of a detail statement or the *approved* construction documents there under, or in violation of a permit or certificate issued under the provisions of this code. Such order shall direct the discontinuance of the illegal action or condition and the abatement of the violation.

[A] 113.3 Prosecution of violation. If the notice of violation is not complied with promptly, the *code official* shall request the legal counsel of the jurisdiction to institute the appropriate proceeding at law or in equity to restrain, correct or abate such violation, or to require the removal or termination of the unlawful pool or spa in violation of the provisions of this code or of the order or direction made pursuant thereto.

[A] 113.4 Violation penalties. Any person who shall violate a provision of this code or shall fail to comply with any of the requirements thereof or who shall erect, install, alter or repair a pool or spa in violation of the *approved* construction documents or directive of the *code official*, or of a permit or certificate issued under the provisions of this code, shall be guilty of a [SPECIFY OFFENSE], punishable by a fine of not more than [AMOUNT] dollars or by imprisonment not exceeding [NUMBER OF DAYS], or both such fine and imprisonment. Each day that a violation continues after due notice has been served shall be deemed a separate offense.

[A] 113.5 Abatement of violation. The imposition of the penalties herein prescribed shall not preclude the legal officer of the jurisdiction from instituting appropriate action to prevent violation, or to prevent illegal use of a pool or spa, or to stop an illegal act, conduct, business or utilization of the plumbing on or about any premises.

[A] 113.6 Unsafe systems. Any system regulated by this code that is unsafe or that constitutes a fire or health hazard, insanitary condition, or is otherwise dangerous to human life is hereby declared unsafe. Any use of a system regulated by this code constituting a hazard to safety, health or public welfare by reason of inadequate maintenance, dilapidation, obsolescence, fire hazard, disaster, damage or abandonment is hereby declared an unsafe use. Any such unsafe system is hereby declared to be a public nuisance and shall be abated by repair, rehabilitation, demolition or removal.

[A] 113.6.1 Authority to condemn a system. Where the *code official* determines that any system, or portion thereof, regulated by this code has become hazardous to life, health or property or has become insanitary, the *code official* shall order in writing that such system either be removed or restored to a safe or sanitary condition. A time limit for compliance with such order shall be specified in the written notice. A person shall not use or maintain a defective system after receiving such notice.

Where such a system is to be disconnected, written notice as prescribed in Section 113.2 shall be given. In cases of immediate danger to life or property, such disconnection shall be made immediately without such notice.

[A] 113.6.2 Authority to disconnect service utilities. The *code official* shall have the authority to authorize disconnection of utility service in accordance with Section 110.3.

[A] 113.6.3 Connection after order to disconnect. A person shall not make connections from any energy, fuel, power supply or water distribution system, or supply energy, fuel or water to any equipment regulated by this code that has been disconnected or ordered to be disconnected by the *code official* or the use of which has been ordered to be discontinued by the *code official* until the *code official* authorizes the reconnection and use of such equipment.

When any system is maintained in violation of this code, and in violation of any notice issued pursuant to the provisions of this section, the *code official* shall institute any appropriate action to prevent, restrain, correct or abate the violation.

SECTION 114—STOP WORK ORDER

[A] 114.1 Authority. Where the *code official* finds any work regulated by this code being performed in a manner contrary to the provisions of this code or in a dangerous or unsafe manner, the *code official* is authorized to issue a stop work order.

[A] 114.2 Issuance. The stop work order shall be in writing and shall be given to the owner of the property, the owner's authorized agent or the person performing the work. Upon issuance of a stop work order, the cited work shall immediately cease. The stop work order shall state the reason for the order and the conditions under which the cited work is authorized to resume.

[A] 114.3 Emergencies. Where an emergency exists, the *code official* shall not be required to give a written notice prior to stopping the work.

[A] 114.4 Failure to comply. Any person who shall continue any work after having been served with a stop work order, except such work as that person is directed to perform to remove a violation or unsafe condition, shall be subject to fines established by the authority having jurisdiction.

User notes:

About this chapter: *Codes, by their very nature, are technical documents. Every word, term and punctuation mark can add to or change the meaning of a technical requirement. It is necessary to maintain a consensus on the specific meaning of each term contained in the code. Chapter 2 performs this function by stating clearly what specific terms mean for the purpose of the code.*

SECTION 201—GENERAL

201.1 Scope. Unless otherwise expressly stated, the following words and terms shall, for the purposes of this code, have the meanings shown in this chapter.

201.2 Interchangeability. Words used in the present tense include the future; words stated in the masculine gender include the feminine and neuter; the singular number includes the plural and the plural, the singular.

201.3 Terms defined in other codes. Where terms are not defined in this code and are defined in the *International Building Code, International Energy Conservation Code, International Fire Code, International Fuel Gas Code, International Mechanical Code, International Plumbing Code* or *International Residential Code,* such terms shall have the meanings ascribed to them as in those codes.

201.4 Terms not defined. Where terms are not defined through the methods authorized by this section, such terms shall have ordinarily accepted meanings such as the context implies.

SECTION 202—DEFINITIONS

ACCESS (TO). That which enables a device, appliance or equipment to be reached by ready access or by a means that first requires the removal or movement of a panel or similar obstruction [see "Ready access (to)"].

ACTIVITY POOL. A pool designed primarily for play activity that uses constructed features and devices including lily pad walks, flotation devices, small slide features, and similar attractions.

AIR INDUCTION SYSTEM. A system whereby a volume of air is introduced into hollow ducting built into a spa floor, bench, or hydrotherapy jets.

[A] ALTERATION. Any construction or renovation to an existing pool or spa other than repair.

[A] APPROVED. Acceptable to the *code official* or authority having jurisdiction.

[A] APPROVED AGENCY. An established and recognized organization regularly engaged in conducting tests or furnishing inspection services, or furnishing product evaluation or certification where such organization has been *approved* by the *code official.*

AQUATIC RECREATION FACILITY. A facility that is designed for free-form aquatic play and recreation. The facilities may include, but are not limited to, wave or surf action pools, leisure rivers, sand bottom pools, vortex pools, activity pools, inner tube rides, body slides and interactive play attractions.

AQUATIC VENUE. A constructed structure or modified natural structure containing water and intended for recreational or therapeutic use. Exposure to water in these structures may occur by contact, ingestion or aerosolization. Examples include swimming pools, wave pools, lazy rivers, surf pools, spas, hot tubs, therapy pools, spray pads, waterpark pools and other interactive water venues.

BACKWASH. The process of cleansing the filter medium or elements by the reverse flow of water through the filter.

BACKWASH CYCLE. The time required to backwash the filter medium or elements and to remove debris in the pool or spa filter.

BARRIER. A permanent fence, wall, building wall, or combination thereof that completely surrounds the pool or spa and obstructs the access to the pool or spa. The term "permanent" shall mean not being able to be removed, lifted, or relocated without the use of a tool.

BATHER. A person using a pool, spa or hot tub and adjoining deck area for the purpose of water sports, recreation, therapy or related activities.

BATHER LOAD. The number of persons in the pool or spa water at any given moment or during any stated period of time.

BEACH ENTRY. Sloping entry starting above the waterline at deck level and ending below the waterline. The presence of sand is not required. Also called "zero entry."

CHEMICAL FEEDER. A floating or mechanical device for adding a chemical to pool or spa water.

CHEMICAL STORAGE SPACE. A space in an aquatic facility used for the storage of pool or spa chemicals such as acids, salt, or corrosive or oxidizing chemicals.

CIRCULATION EQUIPMENT. The components of a circulation system.

CIRCULATION SYSTEM. The mechanical components that are a part of a recirculation system on a pool or spa. Circulation equipment may be, but is not limited to, categories of pumps, hair and lint strainers, filters, valves, gauges, meters, heaters, surface skimmers, inlet fittings, outlet fittings and chemical feeding devices. The components have separate functions, but where connected to each other by piping, perform as a coordinated system for purposes of maintaining pool or spa water in a clear and sanitary condition.

[A] CODE OFFICIAL. The officer or other designated authority charged with the administration and enforcement of this code, or a duly authorized representative.

[A] CONSTRUCTION DOCUMENTS. Written, graphic and pictorial documents prepared or assembled for describing the design, location and physical characteristics of the elements of a project necessary for obtaining a building permit.

COPPER ALLOY. A homogeneous mixture of two or more metals in which copper is the primary component, such as brass and bronze.

DECK. An area immediately adjacent to or attached to a pool or spa that is specifically constructed or installed for sitting, standing, or walking.

DEEP AREA. Water depth areas exceeding 5 feet (1524 mm).

DESIGN PROFESSIONAL. An individual who is registered or licensed to practice his or her respective design profession as defined by the statutory requirements of the professional registration or licensing laws of the state or jurisdiction in which the project is to be constructed.

DESIGN RATE OF FLOW. The rate of flow used for design calculations in a system.

DESIGN WATERLINE. The centerline of the *skimmer* or other point as defined by the designer of the pool or spa.

DIVING AREA. The area of a swimming pool that is designed for diving.

DIVING BOARD. A flexible board secured at one end that is used for diving such as a spring board or a jump board.

DIVING PLATFORM. A stationary platform designed for diving.

DIVING STAND. Any supporting device for a springboard, jump board or diving board.

ELEVATED POOL. Any permanently installed pool, spa, cold plunge, catch basin, overflow trough, including any connected water feature, or body of water feature, that is over a habitable, occupiable or unoccupied space that is (1) inside a thermal envelope, (2) outside a thermal envelope, or (3) a combination of inside and outside the thermal envelope.

EQUIPMENT ROOM. A space intended for the operation of pool pumps, filters, heaters, and controllers. This space is not intended for the storage of hazardous pool or spa chemicals.

EXERCISE SPA (Also known as a swim spa). Variants of a spa in which the design and construction includes specific features and equipment to produce a water flow intended to allow recreational physical activity including, but not limited to, swimming in place. Exercise spas can include peripheral jetted seats intended for water therapy, heater, circulation and filtration system, or can be a separate distinct portion of a combination spa/exercise spa and can have separate controls. These spas are of a design and size such that they have an unobstructed volume of water large enough to allow the 99th Percentile Man as specified in APSP 16 to swim or exercise in place.

EXISTING POOL OR SPA. A pool or spa constructed prior to the date of adoption of this code, or one for which a legal building permit has been issued.

FILTER. A device that removes undissolved particles from water by recirculating the water through a porous substance such as filter medium or elements.

FILTRATION. The process of removing undissolved particles from water by recirculating the water through a porous substance such as filter medium or elements.

[BS] FLOOD HAZARD AREA. The greater of the following two areas:

1. The area within a flood plain subject to a 1-percent or greater chance of flooding in any year.
2. The area designated as a *flood hazard area* on a community's flood hazard map, or otherwise legally designated.

FLUME. A trough-like or tubular structure, generally recognized as a water slide, that directs the path of travel and the rate of descent by the rider.

GUTTER. Overflow trough in the perimeter wall of a pool that is a component of the circulation system or flows to waste.

HAIR AND LINT STRAINER. A device attached on or in front of a pump to which the influent line (suction line) is connected for the purpose of entrapping lint, hair, or other debris that could damage the pump.

HANDHOLD. That portion of a pool or spa structure or a specific element that is at or above the design waterline that users in the pool grasp onto for support.

HANDRAIL. A support device that is intended to be gripped by a user for the purpose of resting or steadying, typically located within or at exits to the pool or spa or as part of a set of steps.

HYDROTHERAPY JET. A fitting that blends air and water, creating a high-velocity turbulent stream of air-enriched water.

INCREASED RISK AQUATIC VENUE. An aquatic venue that has an increased risk of microbial contamination due to its primary users being children under the age of 5 or people more susceptible to infection, such as therapy patients with open wounds. Examples of increased risk aquatic venues include spray pads, wading pools, therapy pools, and other aquatic venues designed primarily for children under the age of 5.

INDOOR AQUATIC FACILITY. A physical place that contains one or more pools or spas and the surrounding bather and spectator/stadium seating areas within a structure that meets the definition of "Building" in the *International Building Code*. It does not include equipment, chemical storage, or bather hygiene rooms or any other rooms with a direct opening to the aquatic facility. Also known as a natatorium.

INTERACTIVE WATER PLAY FEATURES. Any indoor or outdoor structure designed to allow for public recreational activities with recirculated, filtered, and treated water that includes sprayed, jetted or other water sources contacting bathers and not incorporating standing or captured water as part of the bather activity area. These installations are also known as splash pads, spray pads, and wet decks.

JUMP BOARD. A manufactured diving board that has a coil spring, leaf spring, or comparable device located beneath the board that is activated by the force exerted by jumping on the board's end.

[A] JURISDICTION. The governmental unit that has adopted this code.

[A] LABEL. An identification applied on a product by the manufacturer that contains the name of the manufacturer, the function and performance characteristics of the product or material, and the name and identification of an *approved* agency and that indicates that the representative sample of the product or material has been tested and evaluated by an *approved* agency.

[A] LABELED. Equipment, materials or products to which has been affixed a label, seal, symbol or other identifying mark of a nationally recognized testing laboratory, *approved* agency or other organization concerned with product evaluation that maintains periodic inspection of the production of the above-labeled items and whose *labeling* indicates either that the equipment, material or product meets identified standards or has been tested and found suitable for a specified purpose.

LADDER. A structure for ingress and egress that usually consists of two long parallel side pieces joined at intervals by crosspieces such as treads.

> **Type A double access ladder.** An "A-Frame" ladder that straddles the pool wall of an above-ground pool and provides ingress and egress and is intended to be removed when not in use.

> **Type B limited access ladder.** An "A-Frame" ladder that straddles the pool wall of an above-ground/onground pool. Type B ladders are removable and have a built-in feature that prevents entry to the pool when the pool is not in use.

> **Type C ladder.** A "ground to deck" staircase ladder that allows access to an above-ground pool deck and has a built-in entry-limiting feature.

> **Type D in-pool ladder.** Located in the pool to provide a means of ingress and egress from the pool to the deck.

> **Type E or F in-pool staircase ladder.** Located in the pool to provide a means of ingress and egress from the pool to the deck.

LIFELINE. An anchored line thrown to aid in rescue.

[A] LISTED. Equipment, materials, products or services included in a list published by an organization acceptable to the code official and concerned with evaluation of products or services that maintains periodic inspection of production of *listed* equipment or materials or periodic evaluation of services and whose listing states either that the equipment, material, product or service meets identified standards or has been tested and found suitable for a specified purpose. Terms that are used to identify listed equipment, products, or materials include "listed," "certified," "classified" or other terms as determined appropriate by the listing organization.

MAINTAINED ILLUMINATION. The value, in foot-candles or equivalent units, below which the average illuminance on a specified surface is not allowed to fall. *Maintained illumination* equals the initial average illuminance on the specified surface with new lamps, multiplied by the light loss factor (LLF), to account for reduction in lamp intensity over time.

NEGATIVE EDGE. See "*Vanishing edge.*"

NONENTRY AREA. An area of the deck from which entry into the pool or spa is prohibited.

ONGROUND STORABLE POOL. A pool that can be disassembled for storage or transport. This includes portable pools with flexible or nonrigid walls that achieve their structural integrity by means of uniform shape, a support frame or a combination thereof, and that can be disassembled for storage or relocation.

OVERFLOW GUTTER. The gutter around the top perimeter of the pool or spa, which is used to skim the surface.

[A] OWNER. Any person, agent, operator, entity, firm or corporation having any legal or equitable interest in the property; or recorded in the official records of the state, county or municipality as holding an interest or title to the property; or otherwise having possession or control of the property, including the guardian of the estate of any such person, and the executor or administrator of the estate of such person if ordered to take possession of real property by a court.

PEER REVIEW. An independent and objective technical review conducted by an approved third party

PERIMETER FLOW POOL. A pool where the water surface is lifted and flows over the perimeter of the pool into a surrounding gutter that delivers water to the circulation pump.

[A] PERMIT. An official document or certificate issued by the authority having jurisdiction that authorizes performance of a specified activity.

POOL. See "*Public swimming pool*" and "*Residential swimming pool.*"

POWER SAFETY COVER. A pool cover that is placed over the water area, and is opened and closed with a motorized mechanism activated by a control switch.

PUBLIC SWIMMING POOL (Public Pool). A pool, other than a *residential* pool, that is intended to be used for swimming or bathing and is operated by an owner, lessee, operator, licensee or concessionaire, regardless of whether a fee is charged for use. Public pools shall be further classified and defined as follows:

Class A competition pool. A pool intended for use for accredited competitive aquatic events such as Federation Internationale De Natation (FINA), USA Swimming, USA Diving, USA Synchronized Swimming, USA Water Polo, National Collegiate Athletic Association (NCAA), or the National Federation of State High School Associations (NFHS).

Class B public pool. A pool intended for public recreational use that is not identified in the other classifications of public pools.

Class C semi-public pool. A pool operated solely for and in conjunction with lodgings such as hotels, motels, apartments or condominiums.

Class D-1 wave action pool. A pool designed to simulate breaking or cyclic waves for purposes of general play or surfing.

Class D-2 activity pool. A pool designed for casual water play ranging from simple splashing activity to the use of attractions placed in the pool for recreation.

Class D-3 catch pool. A body of water located at the termination of a manufactured waterslide attraction. The body of water is provided for the purpose of terminating the slide action and providing a means for exit to a deck or walkway area.

Class D-4 leisure river. A manufactured stream of water of near-constant depth in which the water is moved by pumps or other means of propulsion to provide a river-like flow that transports bathers over a defined path that may include water features and play devices.

Class D-5 vortex pool. A circular pool equipped with a method of transporting water in the pool for the purpose of propelling riders at speeds dictated by the velocity of the moving stream of water.

Class D-6 interactive play attraction. A manufactured water play device or a combination of water-based play devices in which water flow volumes, pressures or patterns can be varied by the bather without negatively influencing the hydraulic conditions for other connected devices. These attractions incorporate devices or activities such as slides, climbing and crawling structures, visual effects, user-actuated mechanical devices and other elements of bather-driven and bather-controlled play.

Class E. Pools used for instruction, play or therapy and with temperatures above 86°F (30°C).

Class F. Class F pools are wading pools and are covered within the scope of this code as set forth in Section 405.

Public pools are either a diving or nondiving type. Diving types of public pools are classified into types as an indication of the suitability of a pool for use with diving equipment.

Type 0. A nondiving public pool.

Types VI–IX. Public pools suitable for the installation of diving equipment by type.

READY ACCESS (TO). That which enables a device, appliance or equipment to be directly reached, without requiring the removal or movement of any panel or similar obstruction [see "Access (to)"].

RECESSED TREADS. A series of vertically spaced cavities in a pool or spa wall creating tread areas for step holes.

RECIRCULATION SYSTEM. See *"Circulation system."*

REGISTERED DESIGN PROFESSIONAL. An architect or engineer, registered or licensed to practice professional architecture or engineering, as defined by the statutory requirements of the professional registration laws of the state in which the project is to be constructed.

[A] REPAIR. The reconstruction or renewal of any part of a pool or spa for the purpose of its maintenance or to correct damage.

RESIDENTIAL. For purposes of this code, *residential* applies to detached one- and two-family dwellings and townhouses not more than three stories in height.

RESIDENTIAL SWIMMING POOL (Residential Pool). A pool intended for use that is accessory to a *residential* setting and available only to the household and its guests. Other pools shall be considered to be public pools for purposes of this code.

Type 0. A nondiving *residential* pool.

Types I–V. *Residential* pools suitable for the installation of diving equipment by type.

RETURN INLET. The aperture or fitting through which the water under positive pressure returns into a pool.

RING BUOY. A ring-shaped floating buoy capable of supporting a user, usually attached to a throwing line.

ROPE AND FLOAT LINE. A continuous line not less than $^1/_4$ inch (6 mm) in diameter that is supported by buoys and attached to opposite sides of a pool to separate the deep and shallow ends.

RUNOUT. A continuation of water slide flume surface where riders are intended to decelerate and come to a stop.

SAFETY COVER. A structure, fabric or assembly, along with attendant appurtenances and anchoring mechanisms, that is temporarily placed or installed over an entire pool, spa or hot tub and secured in place after all bathers are absent from the water.

SECONDARY DISINFECTION SYSTEM. Disinfection processes or systems installed in increased-risk aquatic venues in addition to the required primary disinfection system.

SHALL. The term, where used in the code, is construed as mandatory.

SHALLOW AREAS. Portions of a pool or spa with water depths less than or equal to 5 feet (1524 mm).

SHOTCRETE. Concrete, wet or dry, placed by a high-velocity pneumatic projection from a nozzle.

SKIMMER. A device installed in the pool or spa that permits the removal of floating debris and surface water to the filter.

SLIP RESISTANT. A surface that has been treated or constructed to significantly reduce the chance of a user slipping. The surface shall not be an abrasion hazard.

SLOPE BREAK. Occurs at the point where the slope of the pool floor changes to a greater slope.

SPA. A product intended for the immersion of persons in temperature-controlled water circulated in a closed system, and not intended to be drained and filled with each use. A spa usually includes a filter, an electric, solar or gas heater, a pump or pumps, and a control, and can include other equipment, such as lights, blowers, and water-sanitizing equipment.

    Nonself-contained spa. A factory-built *spa* in which the water heating and circulating equipment is not an integral part of the product. Nonself-contained spas may employ separate components such as an individual filter, pump, heater and controls, or they can employ assembled combinations of various components.

    Permanent residential spa. A spa, intended for use that is accessory to a *residential* setting and available to the household and its guests and where the water heating and water-circulating equipment is not an integral part of the product. The spa is intended as a permanent plumbing fixture and not intended to be moved.

    Portable residential spa. A spa intended for use that is accessory to a *residential* setting and available to the household and its guests and where it is either self-contained or nonself-contained.

    Public spa. A spa other than a permanent *residential* spa or portable *residential* spa that is intended to be used for bathing and is operated by an owner, licensee or concessionaire, regardless of whether a fee is charged for use.

    Self-contained spa. A factory-built spa in which all control, water heating and water-circulating equipment is an integral part of the product. Self-contained spas may be permanently wired or cord connected.

SPRAY POOL. A pool or basin occupied by construction features that spray water in various arrays for the purpose of wetting the persons playing in the spray streams.

SUBMERGED VACUUM FITTING. A fitting intended to provide a point of connection for suction side automatic swimming pool, *spa*, and hot tub cleaners.

SUCTION OUTLET. Any appurtenance that provides a localized low-pressure area for the transfer of water from a pool to an individual suction system including but not limited to a suction outlet fitting assembly, skimmer or vacuum port fitting.

SUCTION OUTLET FITTING ASSEMBLY (SOFA). A fully submerged suction outlet composed of all components, including the cover and/or grate, adapters, supports, riser rings, a field-built sump or manufactured sump, and fasteners.

SUN SHELF. An area of a pool that adjoins the pool wall with a water depth less than 12 inches (305 mm) and is used for seating and play.

SURFACE SKIMMING SYSTEM. A device or system installed in the pool or spa that permits the removal of floating debris and surface water to the filter.

SURGE CAPACITY. The storage volume in a surge tank, *gutter,* and plumbing lines.

SURGE TANK. A storage vessel within the pool recirculating system used to contain the water displaced by bathers.

SWIMOUT. An underwater seat area that is placed completely outside of the diving envelope of the pool. Where located at the deep end, swimouts are permitted to be used as the deep-end means of entry or exit to the pool.

TUBE RIDE. A gravity flow attraction found at a waterpark designed to convey riders on an inner-tube-like device through a series of chutes, channels, flumes or pools.

TURNOVER RATE. The period of time, usually in hours, required to circulate a volume of water equal to the pool or spa capacity.

UNDERWATER BENCH. An underwater seat that can be recessed into the pool wall or placed completely inside the perimeter shape of the pool, such as a sun shelf.

UNDERWATER LEDGE. A narrow shelf projecting from the side of a vertical structure whose dimensions are defined in the appropriate standard.

VANISHING EDGE. Water-feature detail in which water flows over the edge of not fewer than one of the pool walls and is collected in a catch basin. Also called "Negative edge."

WATERLINE. See "*Design waterline.*"

WAVE POOL CAISSON. A large chamber used in wave generation. This chamber houses pulsing water and air surges in the wave generation process and is not meant for human occupancy.

ZERO ENTRY. See "*Beach entry.*"

GENERAL COMPLIANCE

User notes:

About this chapter: *Chapter 3 covers general regulations for pool and spa installations. As many of these requirements would need to be repeated in Chapters 3 through 10, placing such requirements in only one location eliminates code development coordination issues with the same requirement in multiple locations. These general requirements can be superseded by more specific requirements for certain applications in Chapters 3 through 10.*

SECTION 301—GENERAL

301.1 Scope. The provisions of this chapter shall govern the general design and construction of public and *residential* pools and spas and related piping, equipment, and materials. Provisions that are unique to a specific type of pool or spa are located in Chapters 4 through 10.

   301.1.1 Application of Chapters 4 through 10. Where differences occur between the provisions of this chapter and the provisions of Chapters 4 through 10, the provisions of Chapters 4 through 10 shall apply.

SECTION 302—ELECTRICAL, PLUMBING, MECHANICAL AND FUEL GAS REQUIREMENTS

302.1 Electrical. Electrical requirements for aquatic facilities shall be in accordance with NFPA 70 or the *International Residential Code*, as applicable in accordance with Section 102.7.1.

   Exception: Internal wiring for portable *residential* spas and portable *residential* exercise spas *listed* and *labeled* in accordance with UL 1563 or CSA C22.2 No. 218.1.

302.2 Water service and drainage. Piping and fittings used for water service, makeup and drainage piping for pools and spas shall comply with the *International Plumbing Code*. Fittings shall be *approved* for installation with the piping installed.

302.3 Pipe, fittings and components. Pipe, fittings and components shall be *listed* and *labeled* in accordance with NSF 50 or NSF 14. Plastic jets, fittings, and outlets used in public spas shall be *listed* and *labeled* in accordance with NSF 50.

   Exceptions:
   1. Portable *residential* spas and portable *residential* exercise spas *listed* and *labeled* in accordance with UL 1563 or CSA C22.2 No. 218.1.
   2. *Onground storable pools* supplied by the pool manufacturer as a kit that includes all pipe, fittings and components.

   302.3.1 Suction outlet fitting assembly sumps. Sumps shall be inspected for dimensional conformance to APSP 16 as specified by the suction outlet fitting assembly installation instructions.

302.4 Concealed piping inspection. Piping, including process piping, that is installed in trenches, shall be inspected prior to backfilling.

302.5 Backflow protection. Water supplies for pools and spas shall be protected against backflow in accordance with the *International Plumbing Code* or the *International Residential Code*, as applicable in accordance with Section 102.7.1.

302.6 Wastewater discharge. Where wastewater from pools or spas, such as backwash water from filters and water from deck drains discharge to a building drainage system, the connection shall be through an air gap in accordance with the *International Plumbing Code* or the *International Residential Code* as applicable in accordance with Section 102.7.1.

302.7 Tests. Tests on water piping systems constructed of plastic piping shall not use compressed air for the test.

302.8 Maintenance. Pools and spas shall be maintained in a clean and sanitary condition, and in good repair.

   302.8.1 Manuals. An operating and maintenance manual in accordance with industry-accepted standards shall be provided for each piece of equipment requiring maintenance.

SECTION 303—ENERGY

303.1 Energy consumption of pools and permanent spas. The energy consumption of pools and permanent spas shall be controlled by the requirements in Sections 303.1.1 through 303.1.3.

   303.1.1 Heaters. The electric power to heaters shall be controlled by   on-off switch with ready access that is an integral part of the heater, mounted on the exterior of the heater or external to and within 3 feet (914 mm) of the heater. Operation of such switch shall not change the setting of the heater thermostat. Such switches shall be in addition to a circuit breaker for the power to the heater. Gas-fired heaters shall not be equipped with continuously burning ignition pilots.

303.1.2 Time switches. Time switches or other control methods that can automatically turn off and on heaters and pump motors according to a preset schedule shall be installed for heaters and pump motors. Heaters and pump motors that have built-in time switches shall be in compliance with this section.

Exceptions:

1. Where public health standards require 24-hour pump operation.
2. Pumps that operate solar- or waste-heat recovery pool heating systems.

303.1.3 Covers. Outdoor heated pools and outdoor permanent spas shall be provided with a vapor-retardant cover or other *approved* vapor-retardant means in accordance with Section 104.9.1.

Exception: Where more than 75 percent of the energy for heating, computed over an operating season of not fewer than 3 calendar months, is from a heat pump or an on-site renewable energy system, covers or other vapor-retardant means shall not be required.

303.2 Portable spas. The energy consumption of electric-powered portable spas shall be controlled by the requirements of APSP 14.

303.3 Residential pools and permanent residential spas. The energy consumption of *residential* swimming pools and permanent *residential* spas shall be controlled in accordance with the requirements of APSP 15.

SECTION 304—FLOOD HAZARD AREAS

304.1 General. The provisions of Section 304 shall control the design and construction of pools and spas installed in *flood hazard areas.*

[BS] 304.2 Determination of impacts based on location. Pools and spas located in *flood hazard areas* indicated within the *International Building Code* or the *International Residential Code* shall comply with Section 304.2.1 or 304.2.2.

Exception: Pools and spas located in riverine *flood hazard areas* that are outside of designated floodways and pools and spas located in *flood hazard areas* where the source of flooding is tides, storm surges or coastal storms.

[BS] 304.2.1 Pools and spas located in designated floodways. Where pools and spas are located in designated floodways, documentation shall be submitted to the *code official* that demonstrates that the construction of the pools and spas will not increase the design flood elevation at any point within the jurisdiction.

[BS] 304.2.2 Pools and spas located where floodways have not been designated. Where pools and spas are located where design flood elevations are specified but floodways have not been designated, the applicant shall provide a floodway analysis that demonstrates that the proposed pool or spa and any associated grading and filling, will not increase the design flood elevation more than 1 foot (305 mm) at any point within the jurisdiction.

[BS] 304.3 Pools and spas in coastal high-hazard areas *and coastal A zones*. Pools and spas installed in coastal high-hazard areas and coastal A zones shall be designed and constructed in accordance with ASCE 24.

[BS] 304.4 Protection of equipment. Equipment shall be elevated to or above the design flood elevation.

Exception: Equipment for pools, spas and water features shall be permitted below the required elevation provided that the equipment is elevated to the highest extent practical, is anchored to prevent flotation and resist flood forces, and is protected to prevent water from entering or accumulating within the components during conditions of flooding.

304.5 GFCI protection. Electrical equipment installed below the design flood elevation shall be supplied by branch circuits that have ground-fault circuit interrupter protection for personnel.

SECTION 305—BARRIER REQUIREMENTS

305.1 General. The provisions of this section shall apply to the design of barriers for restricting entry into areas having pools and spas. Where spas or hot tubs are equipped with a lockable *safety cover* complying with ASTM F1346 and swimming pools are equipped with a powered *safety cover* that complies with ASTM F1346, the areas where those spas, hot tubs or pools are located shall not be required to comply with Sections 305.2 through 305.7.

305.1.1 Construction fencing required. The construction sites for in-ground swimming pools and spas shall be provided with construction fencing to surround the site from the time that any excavation occurs up to the time that the permanent barrier is completed. The fencing shall be not less than 4 feet (1219 mm) in height.

305.2 Outdoor swimming pools and spas. Outdoor pools and spas and indoor swimming pools shall be surrounded by a barrier that complies with Sections 305.2.1 through 305.7.

305.2.1 Barrier height and clearances. Barrier heights and clearances shall be in accordance with all of the following:

1. The top of the barrier shall be not less than 48 inches (1219 mm) above grade where measured on the side of the barrier that faces away from the pool or spa. Such height shall exist around the entire perimeter of the barrier and for a distance of 3 feet (914 mm) measured horizontally from the outside of the required barrier.
2. The vertical clearance between grade and the bottom of the barrier shall not exceed 2 inches (51 mm) for grade surfaces that are not solid, such as grass or gravel, where measured on the side of the barrier that faces away from the pool or spa.

3. The vertical clearance between a surface below the barrier to a solid surface, such as concrete, and the bottom of the required barrier shall not exceed 4 inches (102 mm) where measured on the side of the required barrier that faces away from the pool or spa.

4. Where the top of the pool or spa structure is above grade, the barrier shall be installed on grade or shall be mounted on top of the pool or spa structure. Where the barrier is mounted on the top of the pool or spa, the vertical clearance between the top of the pool or spa and the bottom of the barrier shall not exceed 4 inches (102 mm).

305.2.2 Openings. Openings in the barrier shall not allow passage of a 4-inch-diameter (102 mm) sphere.

305.2.3 Solid barrier surfaces. Solid barriers that do not have openings shall not contain indentations or protrusions that form handholds and footholds, except for normal construction tolerances and tooled masonry joints.

305.2.4 Screen enclosure as a barrier. A swimming pool screen enclosure shall be permitted to be utilized as part, or all, of a required barrier provided that the enclosure complies with the requirements of Section 305.2. Such screen enclosures shall be designed by a registered design professional. Walls of such screen enclosures shall not be considered to be dwelling walls.

305.2.4.1 Mesh for screen enclosures. The mesh utilized in the barrier portion of the screen enclosure shall have a tensile strength of not less than 100 pounds per square foot (20.5 kg/m^2) when tested in accordance with ASTM D5034 and a ball burst strength of not less than 150 pounds per square foot (30.7 kg/m^2) when tested in accordance with ASTM D3787.

305.2.5 Mesh fence as a barrier. Mesh fences, other than chain link fences in accordance with Section 305.2.8, shall be installed in accordance with the manufacturer's instructions and shall comply with ASTM F2286 and with both of the following:

1. Where a hinged gate is used with a mesh fence, the gate shall comply with Section 305.3.
2. Mesh fences shall not be installed on top of onground *residential* pools.

305.2.5.1 Setback for mesh fences. The inside of a mesh fence shall be not closer than 20 inches (508 mm) to the nearest edge of the water of a pool or spa.

305.2.6 Closely spaced horizontal members. Where the barrier is composed of horizontal and vertical members and the distance between the tops of the horizontal members is less than 45 inches (1143 mm), the horizontal members shall be located on the pool or spa side of the fence. Spacing between vertical members shall not exceed $1^3/_4$ inches (44 mm) in width. Where there are decorative cutouts within vertical members, spacing within the cutouts shall not exceed $1^3/_4$ inches (44 mm) in width.

305.2.7 Widely spaced horizontal members. Where the barrier is composed of horizontal and vertical members and the distance between the tops of the horizontal members is 45 inches (1143 mm) or more, spacing between vertical members shall not exceed 4 inches (102 mm). Where there are decorative cutouts within vertical members, the interior width of the cutouts shall not exceed $1^3/_4$ inches (44 mm).

305.2.8 Chain link dimensions. The maximum opening formed by a chain link fence shall be not more than $1^3/_4$ inches (44 mm). Where the fence is provided with slats fastened at the top and bottom that reduce the openings, such openings shall be not greater than $1^3/_4$ inches (44 mm).

305.2.9 Diagonal members. Where the barrier is composed of diagonal members, the maximum opening formed by the diagonal members shall be not greater than $1^3/_4$ inches (44 mm). The angle of diagonal members shall be not greater than 45 degrees (0.79 rad) from vertical.

305.2.10 Clear zone. Where equipment, including pool equipment such as pumps, filters and heaters, is on the same lot as a pool or spa and such equipment is located outside of the barrier protecting the pool or spa, such equipment shall be located not less than 36 inches (914 mm) from the outside of the barrier.

305.3 Doors and gates. Doors and gates in barriers shall comply with the requirements of Sections 305.3.1 through 305.3.3 and shall be equipped to accommodate a locking device. Pedestrian access doors and gates shall open outward away from the pool or spa, shall be self-closing and shall have a self-latching device. Doors and gates shall not swing over stairs.

305.3.1 Utility or service doors and gates. Doors and gates not intended for pedestrian use, such as utility or service doors and gates, shall remain locked when not in use.

305.3.2 Double or multiple doors and gates. Double doors and gates or multiple doors and gates shall have not fewer than one leaf secured in place and the adjacent leaf shall be secured with a self-latching device.

305.3.3 Latch release. For doors and gates in barriers, the door and gate latch release mechanisms shall be in accordance with the following:

1. Where door and gate latch release mechanisms are accessed from the outside of the barrier and are not of the self-locking type, such mechanism shall be located above the finished floor or ground surface in accordance with the following:
    1.1. At public pools and spas, not less than 52 inches (1219 mm) and not greater than 54 inches (1372 mm).
    1.2. At *residential* pools and spas, not less than 54 inches (1372 mm).
2. Where door and gate latch release mechanisms are of the self-locking type such as where the lock is operated by means of a key, an electronic opener or the entry of a combination into an integral combination lock, the lock operation control and the latch release mechanism shall be located above the finished floor or ground surface in accordance with the following:
    2.1. At public pools and spas, not less than 34 inches and not greater than 48 inches (1219 mm).
    2.2. At *residential* pools and spas, at not greater than 54 inches (1372 mm).

3. At private pools, where the only latch release mechanism of a self-latching device for a gate is located on the pool and spa side of the barrier, the release mechanism shall be located at a point that is at least 3 inches (76 mm) below the top of the gate.

305.3.4 Barriers adjacent to latch release mechanisms. Where a latch release mechanism is located on the inside of a barrier, openings in the door, gate and barrier within 18 inches (457 mm) of the latch shall not be greater than $^1/_2$ inch (12.7 mm) in any dimension.

305.4 Structure wall as a barrier. Where a wall of a dwelling or structure serves as part of the barrier and where doors, gates or windows provide direct access to the pool or spa through that wall, one of the following shall be required:

1. Operable windows having a sill height of less than 48 inches (1219 mm) above the indoor finished floor, doors and gates shall have an alarm that produces an audible warning when the window, door or their screens are opened. The alarm shall be *listed* and labeled as a water hazard entrance alarm in accordance with UL 2017.

2. In dwellings not required to be Accessible units, Type A units or Type B units, the operable parts of the alarm deactivation switches shall be located at not less than 54 inches (1372 mm) above the finished floor.

3. In dwellings that are required to be Accessible units, Type A units or Type B units, the operable parts of the alarm deactivation switches shall be located not greater than 54 inches (1372 mm) and not less than 48 inches (1219 mm) above the finished floor.

4. In structures other than dwellings, the operable parts of the alarm deactivation switches shall be located not greater than 54 inches (1372 mm) and not less than 48 inches (1220 mm) above the finished floor.

5. A *safety cover* that is *listed* and *labeled* in accordance with ASTM F1346 is installed for the pools and spas.

6. An *approved* means of protection, such as self-closing doors with self-latching devices, is provided. Such means of protection shall provide a degree of protection that is not less than the protection afforded by Item 1 or 2.

305.5 Onground residential pool structure as a barrier. An onground *residential* pool wall structure or a barrier mounted on top of an onground *residential* pool wall structure shall serve as a barrier where all of the following conditions are present:

1. Where only the pool wall serves as the barrier, the bottom of the wall is on grade, the top of the wall is not less than 48 inches (1219 mm) above grade for the entire perimeter of the pool, the wall complies with the requirements of Section 305.2 and the pool manufacturer allows the wall to serve as a barrier.

2. Where a barrier is mounted on top of the pool wall, the top of the barrier is not less than 48 inches (1219 mm) above grade for the entire perimeter of the pool, and the wall and the barrier on top of the wall comply with the requirements of Section 305.2.

3. Ladders or steps used as means of access to the pool are capable of being secured, locked or removed to prevent access except where the ladder or steps are surrounded by a barrier that meets the requirements of Section 305.

4. Openings created by the securing, locking or removal of ladders and steps do not allow the passage of a 4-inch (102 mm) diameter sphere.

5. Barriers that are mounted on top of onground *residential* pool walls are installed in accordance with the pool manufacturer's instructions.

305.6 Natural barriers. In the case where the pool or spa area abuts the edge of a lake or other natural body of water, public access is not permitted or allowed along the shoreline, and required barriers extend to and beyond the water's edge not less than 18 inches (457 mm), a barrier is not required between the natural body of water shoreline and the pool or spa.

305.7 Natural topography. Natural topography that prevents direct access to the pool or spa area shall include but not be limited to mountains and natural rock formations. A natural barrier *approved* by the governing body shall be acceptable provided that the degree of protection is not less than the protection afforded by the requirements of Sections 305.2 through 305.5.

305.8 Means of egress. Outdoor public pools provided with barriers shall have means of egress as required by Chapter 10 of the *International Building Code*.

SECTION 306—DECKS

Scan for Changes

8dc80d5

306.1 General. The structural design and installation of decks around pools and spas shall be in accordance with the *International Residential Code* or the *International Building Code*, as applicable in accordance with Section 102.7 and this section.

306.2 Slip resistant. Decks, ramps, coping, and similar step surfaces shall be slip resistant and cleanable. Where surfaces are evaluated for slip resistance, such surfaces shall have, when tested wet, a minimum pendulum slip rating classification of P4 if tested in accordance with SA AS4586 or a minimum Dynamic Coefficient of Friction (DCOF) of 0.42 if tested in accordance with ANSI A326. The design professional shall determine the appropriate classification and level of slip resistance necessary based on surface type, placement environment, and pedestrian traffic. Special features in or on decks such as markers, brand insignias, and similar materials shall be slip resistant.

306.3 Step risers and treads. Step risers for decks of public pools and spas shall be uniform and have a height not less than $3^3/_4$ inches (95 mm) and not greater than $7^1/_2$ inches (191 mm). The tread distance from front to back shall be not less than 11 inches (279 mm). Step risers and treads for decks of *residential* pools and spas shall be in accordance with the *International Residential Code*.

306.4 Deck steps handrail required. Public pool and spa deck steps having three or more risers shall be provided with a handrail.

306.5 Slope. The minimum slope of decks shall be in accordance with Table 306.5. The maximum slope of decks shall be not greater than $^1/_2$ inch per foot (1 mm per 24 mm).

Exceptions:

1. The minimum slope of decks in Table 306.5 shall not be required where an alternative drainage method is provided that prevents the accumulation or pooling of water deeper than $^1/_8$ inch (3.2 mm), 20 minutes after the cessation of the addition of water to the deck.
2. The minimum slope of decks in Table 306.5 shall not be required where the decking is gapped in accordance with Section 306.6.

TABLE 306.5—MINIMUM DRAINAGE SLOPES FOR DECK SURFACES

SURFACE	MINIMUM DRAINAGE SLOPE (INCH PER FOOT)
Carpet	$^1/_2$
Exposed aggregate	$^1/_4$
Textured, hand-finished concrete	$^1/_8$
Travertine/brick-set pavers, public pools or spas	$^3/_8$
Travertine/brick-set pavers, residential pools or spas	$^1/_8$
Wood	$^1/_8$
Wood/plastic composite	$^1/_8$
For SI: 1 inch = 25.4 mm, 1 foot = 304.8 mm.	

306.5.1 Deck drainage. Decks shall be sloped to drain away from the pool or toward the deck drains. Where site conditions require, deck drains shall be permitted to be placed at the back side of the pool structure or coping.

306.5.2 Site drainage. Site drainage shall direct all perimeter deck drainage, general site, and roof drainage away from the pool area.

Exception: First 3 feet (914 mm) of decking immediately surrounding perimeter flow pools

306.6 Gaps. Gaps not less than $^1/_8$ inch and not greater than $^1/_2$ inch shall be provided between wood deck boards for drainage. Gaps between manufactured deck boards shall be in accordance with the manufacturer's installation instructions.

Exception: Gaps are not required between wood deck boards installed on decks sloped in accordance with Section R306.5

306.6.1 Maximum gap. The open gap between pool decks and adjoining decks or walkways, including joint material, shall be not greater than $^3/_4$ inch (19.1 mm). The difference in vertical elevation between the pool deck and the adjoining sidewalk shall be not greater than $^1/_4$ inch (6.4 mm).

306.7 Concrete joints. Isolation joints that occur where the pool coping meets the concrete deck shall be watertight.

306.7.1 Joints at coping. Joints that occur where the pool coping meets the concrete deck shall be installed to protect the coping and its mortar bed from damage as a result of the anticipated movement of adjoining deck.

306.7.2 Crack control. Joints in a deck shall be provided to minimize visible cracks outside of the control joints caused by imposed stresses or movement of the slab.

306.7.3 Movement control. Areas where decks join existing concrete work shall be provided with a joint to protect the pool from damage caused by relative movement.

306.8 Deck edges. The edges of decks shall be radiused, tapered, or otherwise designed to eliminate sharp corners.

306.9 Valves under decks. Valves installed in or under decks shall be provided access for operation, service, and maintenance. Where access through the deck walking surface is required, an access cover shall be provided for the opening in the deck. Such access covers shall be slip resistant and secured.

306.9.1 Hose bibbs. Hose bibbs shall be provided for rinsing down the entire deck and shall be installed in accordance with the *International Plumbing Code* or *International Residential Code*, as applicable in accordance with Section 102.7.1, and shall be located not greater than 150 feet (45 720 mm) apart. Water-powered devices, such as water-powered lifts, shall have a dedicated hose bibb water source.

Exception: *Residential* pools and spas shall not be required to have hose bibbs located at 150-foot (45 720 mm) intervals, or have a dedicated hose bibb for water-powered devices.

SECTION 307—GENERAL DESIGN

307.1 General design requirements. Sections 307.1.1 through 307.1.5 shall apply to all pools and spas.

307.1.1 Glazing in hazardous locations. Hazardous locations for glazing shall be as defined in the *International Building Code* or the *International Residential Code*, as applicable in accordance with Section 102.7.1 of this code. Where glazing is determined to be in a hazardous location, the requirements for the glazing shall be in accordance with those codes, as applicable.

307.1.2 Colors and finishes. For other than *residential* pools and *residential* spas, the colors, patterns, or finishes of the pool and spa interiors shall not obscure objects or surfaces within the pool or spa. The interior finish coating floors and walls shall be white or light-colored.

307.1.2.1 Munsell gray scale. Finishes shall be not less than 8.0 on the Munsell gray scale.

Exceptions: The following shall not be required to comply with this section:

1. Competitive lane markings.
2. Floors of dedicated competitive diving wells.
3. Step or bench edge markings.
4. Pools shallower than 24 inches (609.6 mm).
5. Water line tiles.
6. Wave and surf pool depth change indicator tiles.
7. Depth change indicator tiles where a rope and float line is provided.
8. Features such as rock formations, as *approved*.

307.1.3 Designs or logos. Any design or logo on the pool floor or walls shall be such that it will not hinder the detection of a human in distress, algae, sediment, or other objects in the pool.

307.1.4 Roofs or canopies. Roofs or canopies over pools and spas shall be in accordance with the *International Building Code* or *International Residential Code*, as applicable in accordance with Section 102.7.1 and shall be constructed so as to prevent water runoff into the pool or spa.

307.1.5 Accessibility. An accessible route to public pools and spas shall be provided in accordance with the *International Building Code*. Accessibility within public pools and spas shall be provided as required by the accessible recreational facilities provisions of the *International Building Code*. Pool and spa lifts providing an accessible means of entry into the water shall be *listed* and *labeled* in accordance with UL 60335-2-1000 and be installed in accordance with ICC A117.1 and NFPA 70.

307.2 Specific design and material requirements. Sections 307.2.1 through 307.2.4 shall apply to all pools and spas except for *listed* and *labeled* portable *residential* spas, and *listed* and *labeled* portable *residential* exercise spas.

307.2.1 Materials. Pools and spas and appurtenances thereto shall be constructed of materials that are nontoxic to humans and the environment; that are generally or commonly regarded to be impervious and enduring; that will withstand the design stresses; and that will provide a watertight structure with a smooth and easily cleanable surface without cracks or joints, excluding structural joints, or that will provide a watertight structure to which a smooth, easily cleaned surface/finish is applied or attached. Material surfaces that come in contact with the user shall be finished, so that they do not constitute a cutting, pinching, puncturing or abrasion hazard under casual contact and intended use.

307.2.1.1 Beach pools. Clean sand or similar material, where used in a beach pool environment, shall be used over an impervious surface. The sand area shall be designed and controlled so that the circulation system, maintenance, safety, sanitation, and operation of the pool are not adversely affected.

307.2.1.2 Compatibility. Assemblies of different materials shall be chemically and mechanically compatible for their intended use and environment.

307.2.2 Materials and structural design. Pools and spas shall conform to one or more of the standards indicated in Table 307.2.2. The structural design of pools and spas shall be in accordance with the *International Building Code* or the *International Residential Code*, as applicable in accordance with Section 102.7.1 of this code.

Exception: Pools and spas constructed with reinforced concrete or reinforced shotcrete with a minimum compressive strength of 2,500 pounds per square inch (175.8 kg/cm²) as designed by a design professional and approved shall be permitted.

TABLE 307.2.2—RESERVOIRS AND SHELLS	
MATERIAL	**STANDARD**
Fiberglass reinforced plastic	IAPMO Z124.7
Plastic	IAPMO Z124.7
Reinforced concrete	ACI 318
Reinforced shotcrete	ACI 318
Stainless steel (Types 316, 316L, 304, 304L)	ASTM A240
Tile	ANSI A108/A118/A136.1

TABLE 307.2.2—RESERVOIRS AND SHELLS—continued

MATERIAL	STANDARD
Vinyl	ASTM D1593

307.2.2.1 Installation. Equipment for pools and spas shall be supported to prevent damage from misalignment and settling and located so as to allow access for inspection, servicing, removal and repair of component parts.

307.2.3 Freeze protection. In climates subject to freezing temperatures, outdoor pool and spa shells and appurtenances, piping, filter systems, pumps and motors, and other components shall be designed and constructed to provide protection from damage from freezing.

307.2.4 Surface condition. The surfaces within public pools and spas intended to provide footing for users shall be slip resistant and shall not cause injury during normal use.

307.2.5 Plaster. The plastering of pools and permanently installed concrete spas shall be in accordance with APSP/NPC/ICC-12.

SECTION 308—ELEVATED POOLS

308.1 Design of elevated pools. Elevated pools shall be designed and constructed in accordance with PHTA 10.

SECTION 309—DIMENSIONAL DESIGN

309.1 Floor slope. The slope of the floor from the point of the first slope change to the deep area shall not exceed one unit vertical in three units horizontal (33-percent slope).

Exception: Portable *residential* spas and portable *residential* exercise spas.

309.2 Walls. Walls shall intersect with the floor at an angle or a transition profile. Where a transitional profile is provided at water depths of 3 feet (914 mm) or less, a transitional radius shall not exceed 6 inches (152 mm) and shall be tangent to the wall and is permitted to be tangent to or intersect the floor.

Exceptions:

1. Portable *residential* spas and portable *residential* exercise spas.
2. *Onground storable pools*.

309.3 Shape. This code is not intended to regulate the shape of a pool or spa other than to take into account the effect that a given shape will have on the safety of the occupants and to maintain the minimum required level of circulation to ensure sanitation.

309.4 Waterline. The *design waterline* shall have a maximum construction tolerance at the time of completion of the work of plus or minus $1/4$ inch (6.4 mm) for pools and spas with adjustable weir surface skimming systems, and plus or minus $1/8$ inch (3.2 mm) for pools and spas with nonadjustable surface skimming systems.

SECTION 310—EQUIPMENT

310.1 Electrically operated equipment. Electrically operated equipment shall be *listed* and *labeled* in accordance with applicable product standards.

Exception: Portable *residential* spas and portable *residential* exercise spas *listed* and *labeled* in accordance with UL 1563 or CSA C22.2 No. 218.1.

310.2 Treatment and circulation system equipment. Treatment and circulation system equipment for public pools and spas shall be *listed* and *labeled* in accordance with NSF 50 and other applicable standards.

SECTION 311—SUCTION ENTRAPMENT AVOIDANCE

311.1 General. Suction entrapment avoidance for pools and spas shall be provided in accordance with APSP 7 (ANSI/PHTA/ICC 7).

Exceptions:

1. Portable spas and portable exercise spas *listed* and *labeled* in accordance with UL 1563 or CSA C22.2 No. 218.1.
2. Suction entrapment avoidance for wading pools shall be provided in accordance with Section 405.

SECTION 312—CIRCULATION SYSTEMS

312.1 General. The provisions of this section shall apply to circulation systems for pools and spas.

Exceptions:

1. Portable *residential* spas and portable *residential* exercise spas.
2. *Onground storable pools* supplied by the pool manufacturer as a kit that includes circulation system equipment that is in accordance with Section 704.

312.2 System design. A circulation system consisting of pumps, piping, return inlets and outlets, filters, and other necessary equipment shall be provided for the complete circulation of water. Wading pools and spas shall have separate dedicated filtering systems.

Exception: Separate filtering systems are not required for *residential* pools and spas.

312.2.1 Turnover rate. The equipment shall be sized to turn over the volume of water that the pool or spa is capable of containing as specified in this code for the specific installation.

312.2.2 Servicing. Circulation system components that require replacement or servicing shall be provided with access for inspection, repair, or replacement and shall be installed in accordance with the manufacturer's specifications.

312.2.3 Equipment anchorage. Pool and spa equipment and related piping shall be designed and installed in accordance with the manufacturer's instructions.

312.3 Water velocity. The water velocity in suction and return piping shall comply with either Section 312.3.1 or 312.3.2. The water velocity in copper and copper alloy piping shall not exceed 8 fps (2.4 mps). All water velocity calculations shall be based on the design flow rate specified for each recirculation system.

312.3.1 Public pools and spas. For public pools and spas, suction piping water velocity shall not exceed 6 fps (1.8 mps) and return piping water velocity shall not exceed 8 fps (2.4 mps).

312.3.2 Residential pools and spas. For *residential* pools and spas, the water velocity in suction piping and return piping shall not exceed 8 fps (2.4 mps).

312.4 Piping and fittings. Plastic pipe and fittings used in circulation systems shall be nontoxic and shall be able to withstand the design operating pressures and conditions of the pool or spa. Plastic pipe shall be *listed* and *labeled* as complying with NSF 14. Circulation system piping shall be *listed* and *labeled* as complying with one of the standards in Table 312.4.

TABLE 312.4—CIRCULATION SYSTEM PIPE MATERIAL STANDARD

MATERIAL	STANDARD
Acrylonitrile butadiene styrene (ABS) plastic pipe	ASTM D1527
Chlorinated polyvinyl chloride (CPVC) plastic pipe and tubing	ASTM D2846; CSA B137.6
Copper or copper-alloy tubing	ASTM B88; ASTM B447
Polyvinyl chloride (PVC) hose	ASTM D1785; ASTM D2241; ASTM D2672; CSA B137.3
Polyvinyl chloride (PVC) plastic pipe	ASTM D1785; CSA B137.3
Stainless steel pipe, Types 304, 304L, 316, 316L	ASTM A312

312.4.1 Fittings. Fittings used in circulation systems shall be *listed* and *labeled* as complying with one of the standards in Table 312.4.1.

Exceptions:

1. Suction outlet fitting assemblies and manufacturer-provided components that conform to APSP 16.
2. Skimmers and manufacturer-provided components.
3. *Gutter* overflow grates and fittings installed above or outside of the overflow point of the pool or spa.

TABLE 312.4.1—CIRCULATION SYSTEM FITTINGS

MATERIAL	STANDARD
Acrylonitrile butadiene styrene (ABS) plastic pipe	ASTM D1527
Chlorinated polyvinyl chloride (CPVC) plastic pipe and tubing	ASTM D2846; ASTM F437; ASTM F438; ASTM F439; CSA B137.6
Copper or copper-alloy tubing	ASME B16.15
Polyvinyl chloride (PVC) plastic pipe	ASTM D2464; ASTM D2466; ASTM D2467; CSA B137.2; CSA B137.3
Stainless steel pipe, Types 304, 304L, 316, 316L	ASTM A182; ASTM A403

312.4.2 Joints. Joints shall be made in accordance with manufacturer's instructions.

312.4.3 Piping subject to freezing. Piping subject to damage by freezing shall have a uniform slope in one direction and shall be equipped with valves for drainage or shall be capable of being evacuated to remove the water.

312.4.4 Suction outlet fitting assemblies. Suction outlet fitting assemblies shall conform to APSP 16. Manufactured suction outlet fitting assemblies shall be listed and labeled. Suction outlet fitting assemblies other than the manufactured type shall be certified as conforming by a design professional.

312.5 System draining. Equipment shall be designed and fabricated to drain the water from the equipment, together with exposed face piping, by removal of drain plugs, manipulating valves, or by other methods. Drainage shall be in accordance with manufacturer's specifications.

312.6 Pressure or vacuum gauge. Gauges shall be provided on the circulation system for public pools. Gauges shall be provided with ready access.

1. A pressure gauge shall be located downstream of the pump and between the pump and filter.
2. A vacuum gauge shall be located between the pump and filter and upstream of the pump.

312.7 Flow measurement. Public swimming pools and wading pools shall be equipped with a flow-measuring device that indicates the rate of flow through the filter system. The flow rate measuring device shall indicate gallons per minute (lpm) and shall be selected and installed to be accurate within plus or minus 10 percent of actual flow.

312.8 Instructions. Written operation and maintenance instructions shall be provided for the circulation system of public pools.

312.9 Hydrostatic pressure test. Circulation system piping, other than that integrally included in the manufacture of the pool or spa, shall be subjected to a hydrostatic pressure test of 25 pounds per square inch (psi) (172.4 kPa). This pressure shall be held for not less than 15 minutes.

SECTION 313—FILTERS

313.1 General. The provisions of this section apply to filters for pools and spas.

Exceptions:

1. Portable *residential* spas and portable *residential* exercise spas.
2. *Onground storable pools* supplied by the pool manufacturer as a kit that includes a filter that is in accordance with Section 704.

313.2 Design. Filters shall have a flow rating equal to or greater than the design flow rate of the system. Filters shall be installed in accordance with the manufacturer's instructions. Filters shall be designed so that filtration surfaces can be inspected and serviced.

313.3 Internal pressure. For pressure-type filters, a means shall be provided to allow the release of internal pressure.

313.3.1 Air release. Filters incorporating an automatic means of internal air release as the principal means of air release shall have one or more lids that provide a slow and safe release of pressure as a part of the design and shall have a manual air release in addition to an automatic release.

313.3.2 Separation tanks. A separation tank used in conjunction with a filter tank shall have a manual method of air release or a lid that provides for a slow and safe release of pressure as it is opened.

SECTION 314—PUMPS AND MOTORS

314.1 General. The provisions of this section apply to pumps and motors for pools and spas.

Exceptions:

1. Portable *residential* spas and portable *residential* exercise spas.
2. *Onground storable pools* supplied by the pool manufacturer as a kit that includes a pump and motor that is in accordance with Section 704.

314.2 Performance. A pump shall be provided for circulation of the pool water. The pump shall be capable of providing the flow required for filtering the pool water and filter cleaning, if applicable, against the total dynamic head developed by the complete system.

314.3 Intake protection. A cleanable strainer, skimmer basket, or screen shall be provided for pools and spas, upstream or as an integral part of circulation pumps, to remove solids, debris, hair, and lint on pressure filter systems.

314.4 Location. Pumps and motors shall be provided with access for inspection and service in accordance with the manufacturer's specifications.

314.5 Safety. The design, construction, and installation of pumps and component parts shall be in accordance with the manufacturer's specifications.

314.6 Isolation valves. Shutoff valves shall be installed on the suction and discharge sides of pumps that are located below the waterline. Such valves shall be provided with access.

314.7 Emergency shutoff switch. An emergency shutoff switch shall be provided to disconnect power to recirculation and jet system pumps and air blowers. Emergency shutoff switches shall be: provided with access; located within sight of the pool or spa; and located not less than 5 feet (1524 mm) horizontally from the inside walls of the pool or spa.

Exception: *Onground storable pools*, permanent inground *residential* swimming pools, *residential* spas and *residential* water features.

314.8 Motor performance. Motors shall comply with UL 1004-1, UL 1081, CSA C22.2 No. 108 or the relevant motor requirements of UL 1563 or CSA C22.2 No. 218.1, as applicable.

SECTION 315—RETURN AND SUCTION FITTINGS

315.1 General. The provisions of this section apply to return and suction fittings for pools and spas

Exception: Portable *residential* spas and portable *residential* exercise spas.

315.2 Entrapment avoidance. Entrapment avoidance means shall be provided in accordance with Section 311.

315.3 Flow distribution. The suction outlet fitting assemblies, where installed, and the skimming systems shall each be designed to accommodate 100 percent of the circulation turnover rate.

315.3.1 Multiple systems. Where multiple systems are used in a single pool to meet this requirement, each subsystem shall proportionately be designed such that the maximum design flow rates cannot be exceeded during normal operation.

315.4 Return inlets. One return inlet shall be provided for every 300 square feet (27.9 m²) of pool surface area, or fraction thereof.

Exception: *Onground storable pools*.

315.4.1 Design. Return and suction fittings for the circulation system shall be designed so as not to constitute a hazard to the bather.

315.5 Vacuum fittings. Where installed, access shall be provided to *submerged vacuum fittings* and such fittings shall be located not greater than 12 inches (305 mm) below the water level.

SECTION 316—SKIMMERS

316.1 General. The provisions of this section apply to skimmers for pools and spas.

Exceptions:

1. Portable *residential* spas and portable *residential* exercise spas.
2. *Onground storable pools* supplied by the pool manufacturer as a kit that includes a skimming system that is in accordance with Section 704.

316.2 Required. A surface skimming system shall be provided for public pools and spas. Surface skimming systems shall be *listed* and *labeled* in accordance with NSF 50. Either a surface skimming system or perimeter overflow system shall be provided for permanent inground *residential* pools and permanent *residential* spas. Where installed, surface skimming systems shall be designed and constructed to create a skimming action on the pool water surface when the water level in the pool is within operational parameters.

Exceptions:

1. Class D public pools designed in accordance with Chapter 6.
2. Skimmers that are an integral part of a spa that has been *listed* and *labeled* in accordance with UL1563 shall not be required to be *listed* and *labeled* in accordance with NSF 50.

316.2.1 Circulation systems. Public pool circulation systems shall be designed to process not less than 100 percent of the turnover rate through skimmers.

316.3 Skimmer sizing. Where automatic surface skimmers are used as the sole overflow system, not less than one surface skimmer shall be provided for the square foot (square meter) areas, or fractions thereof, indicated in Table 316.3. Skimmers shall be located to maintain effective skimming action.

TABLE 316.3—SKIMMER SIZING TABLE	
POOL OR SPA	**AREA PER SKIMMER (SQ. FT)**
Public pool	500
Residential pool	800
Spas (all types)	150
For SI: 1 square foot = 0.0929 m².	

316.4 Perimeter coverage. Where a perimeter-type surface skimming system is used as the sole surface skimming system, the system shall extend around not less than 50 percent of the pool or spa perimeter.

316.4.1 Surge capacity. Where perimeter surface skimming systems are used, they shall be connected to a circulation system with a system surge capacity of not less than 1 gallon for each square foot (40.7 liters per square meter) of water surface. The capacity of the perimeter overflow system and related piping is permitted to be considered as a portion of the surge capacity.

316.5 Equalizers. Equalizers on skimmers shall be prohibited.

316.6 Hazard. Skimming devices shall be designed and installed so as not to create a hazard to the user.

SECTION 317—HEATERS

317.1 General. The provisions of this section apply to heaters for pools and spas.

Exception: Portable *residential* spas and portable *residential* exercise spas.

317.2 Certification. Heaters and hot water storage tanks shall be *listed* and *labeled* in accordance with the applicable standard indicated in Table 317.2(1). Hot water heating systems and components shall comply with the applicable standard indicated in Table 317.2(2).

TABLE 317.2(1)—WATER HEATERS	
DEVICE	**STANDARD**
Electric water heater	UL 1261, UL 1563 or CSA C22.2 No. 218.1
Gas-fired water heater	ANSI Z21.56/CSA 4.7a
Heat exchanger	AHRI 400
Heat pump water heater	AHRI 1160 and one of the following: CSA C22.2 No. 236, UL 1995, or UL/CSA 60335-2-40

TABLE 317.2(2)—WATER HEATING SYSTEMS AND COMPONENTS	
SYSTEM	**STANDARD**
Solar water heater	ICC/APSP 902/SRCC 400

317.3 Sizing. Heaters shall be sized in accordance with the manufacturer's specifications.

317.4 Installation. Heaters shall be installed in accordance with the manufacturer's specifications and the *International Fuel Gas Code*, *International Mechanical Code*, *International Energy Conservation Code*, NFPA 70 or *International Residential Code*, as applicable in accordance with Section 102.7.1. Solar water heating systems shall be installed in accordance with Section 317.6.

317.4.1 Temperature. A means shall be provided to monitor water temperature.

317.4.2 Access prohibited. For public pools and spas, public access to controls shall be prohibited.

317.5 Heater circulation system. Heater circulation systems shall comply with Sections 317.5.1 and 317.5.2.

317.5.1 Water flow. Water flow through the heater bypass piping, back-siphonage protection, and the use of heat sinks shall be in accordance with the heater manufacturer's specifications.

317.5.2 Pump delay. Where required by the manufacturer, heaters shall be installed with an automatic device that will ensure that the pump continues to run after the heater shuts off for the time period specified by the manufacturer.

317.6 Solar water *heating systems*. Solar water heating systems utilized for pools and spas shall comply with Sections 317.6.1 through 317.6.3.

317.6.1 Installation. Solar thermal water heaters shall be installed in accordance with the *International Mechanical Code* or *International Residential Code*, as applicable in accordance with Section 102.7.1.

317.6.2 Certification of collectors. Solar thermal collectors shall be *listed* and *labeled* in accordance with ICC 901/SRCC 100.

317.6.3 Marking of collectors and modules. Solar thermal collectors and photovoltaic modules shall be permanently marked with the manufacturer's name, model number, and serial number. Such markings shall be located on each collector in a position that is readily viewable after installation.

SECTION 318—AIR BLOWER AND AIR INDUCTION SYSTEM

318.1 General. This section applies to devices and systems that induce or allow air to enter Class A and *residential* pools and spas either by means of a powered pump or passive design.

318.2 Backflow prevention. Air blower systems shall be equipped with backflow protection as specified in UL 1563 or CSA C22.2 No. 218.1.

318.3 Air intake source. Air intake sources shall not induce water, dirt or contaminants.

318.4 Sizing. Air induction systems shall be sized in accordance with the manufacturer's specifications.

318.5 Inspection and service. Air blowers shall be provided with access for inspection and service.

SECTION 319—WATER SUPPLY

319.1 Makeup water. Makeup water to maintain the water level and water used as a vehicle for sanitizers or other chemicals, for pump priming, or for other such additions, shall be from a potable water source.

319.2 Protection of potable water supply. Potable water supply systems shall be designed, installed and maintained so as to prevent contamination from nonpotable liquids, solids or gases being introduced into the potable water supply through cross-connections or other piping connections to the system. Means of protection against backflow in the potable water supply shall be provided through an air gap complying with ASME A112.1.2 or by a backflow prevention assembly in accordance with the *International Residential Code* or the *International Plumbing Code*, as applicable in accordance with Section 102.7.1.

319.3 **Over-the-rim spouts.** Over-the-rim spouts shall be located under a diving board, adjacent to a ladder, or otherwise shielded so as not to create a hazard. The open end of such spouts shall not have sharp edges and shall not protrude more than 2 inches (51 mm) beyond the edge of the pool. The open end shall be separated from the water by an air gap of not less than 1.5 pipe diameters measured from the pipe outlet to the rim.

SECTION 320—SANITIZING EQUIPMENT AND CHEMICAL FEEDERS

320.1 Equipment standards. Sanitizing equipment installed in public pools and spas shall be capable of introducing the quantity of sanitizer necessary to maintain the appropriate levels under all conditions of intended use.

320.2 Chemical feeders. Public pools and spas shall be equipped with chemical feed equipment such as flow-through chemical feeders, electrolytic chemical generators, mechanical chemical feeders, chemical feed pumps, and automatic controllers that are *listed* and *labeled* in compliance with NSF 50. Chemical feed systems shall be installed in accordance with the manufacturer's specifications. Chemical feed pumps shall be wired so that they cannot operate unless there is adequate return flow to disburse the chemical throughout the pool or spa as designed.

320.3 Secondary disinfection systems. Secondary disinfection systems shall be installed for the following increased-risk aquatic venues in addition to the required primary disinfection system:

1. Wading pools.
2. Interactive water play features.
3. Therapy pools.
4. Other aquatic venues designed primarily for children under the age of 5.

The secondary disinfection system shall be *listed* and labeled to NSF 50 and installed in accordance with the manufacturer's specifications. Where electrically powered, such equipment shall additionally be listed and labeled in accordance with UL 1081 or UL 1563.

320.4 Supplemental treatment systems. Supplemental treatment systems in public pools and spas shall be certified to NSF 50 and installed in accordance with the manufacturer's specifications. Where electrically powered, such equipment shall additionally be *listed* and *labeled* in accordance with UL 1081 or UL 1563.

SECTION 321—WASTEWATER DISPOSAL

321.1 Backwash water or draining water. Backwash water and draining water shall be discharged to the sanitary or storm sewer, or into an *approved* disposal system on the premise, or shall be disposed of by other means *approved* by the state or local authority. Direct connections shall not be made between the end of the backwash line and the disposal system. Drains shall discharge through an air gap.

321.2 Water salvage. Filter backwash water shall not be returned to the vessel except where the backwash water has been filtered to remove particulates, treated to eliminate coliform bacteria and waterborne pathogens, and such return has been *approved* by the state or local authority.

321.3 Waste post treatment. Where necessary, filter backwash water and drainage water shall be treated chemically or through the use of settling tanks to eliminate or neutralize chemicals, diatomaceous earth, and contaminants in the water that exceed the limits set by the state or local effluent discharge requirements.

SECTION 322—LIGHTING

322.1 General. The provisions of Sections 322.2 and 322.3 shall apply to lighting for public pools and spas. The provisions of Section 322.4 shall apply to lighting for *residential* pools and spas.

322.2 Artificial lighting required. When a pool is open during periods of low natural illumination, artificial lighting shall be provided so that all areas of the pool, including all suction outlets on the bottom of the pool, will be visible. Illumination shall be sufficient to enable a lifeguard or other persons standing on the deck or sitting on a lifeguard stand adjacent to the pool edge to determine if a pool user is lying on the bottom of the pool and that the pool water is transparent and free from cloudiness.

These two conditions shall be met when all suction outlets are visible from the edge of the deck at all times when artificial lighting is illuminated and when an 8-inch-diameter (152 mm) black disk, placed at the bottom of the pool in the deepest point, is visible from the edge of the pool deck at all times when artificial lighting is illuminated.

322.2.1 Pool and deck illumination. Overhead lighting, underwater lighting or both shall be provided to illuminate the pool and adjacent deck areas. The lighting shall be *listed* and *labeled*. The lighting shall be installed in accordance with NFPA 70.

322.2.2 Illumination intensity. For outdoor pools, any combination of overhead and underwater lighting shall provide *maintained illumination* not less than 10 horizontal foot-candles (10 lumens per square foot) [108 lux] at the pool water surface. For indoor pools, any combination of overhead and underwater lighting shall provide *maintained illumination* of not less than 30 horizontal foot-candles (30 lumens per square foot) [323 lux] at the pool water surface. Deck area lighting for both indoor and outdoor pools shall provide *maintained illumination* of not less than 10 horizontal foot-candles (10 lumens per square foot) [108 lux] at the walking surface of the deck.

322.2.3 Underwater lighting. Underwater lighting shall provide not less than 8 lamp lumens per square foot of pool water surface area.

> **Exception:** The requirement of this section shall not apply where overhead lighting provides not less than 15 foot-candles (15 lumens per square foot) [161 lux] of *maintained illumination* at the pool water surface, the overhead lighting provides visibility, without glare, of all areas of the pool, and the requirements of Section 322.2.2 are met or exceeded.

322.3 Emergency illumination. Public pools and public pool areas that operate during periods of low illumination shall be provided with emergency lighting that will automatically turn on to permit evacuation of the pool and securing of the area in the event of power failure. Emergency lighting facilities shall be arranged to provide initial illumination that is not less than 0.1 foot-candle (0.1 lumen per square foot) [1 lux] measured at any point on the water surface and at any point on the walking surface of the deck, and not less than an average of 1 foot-candle (1 lumen per square foot) [11 lux]. At the end of the emergency lighting time duration, the illumination level shall be not less than 0.06 foot-candle (0.06 lumen per square foot) [0.65 lux] measured at any point on the water surface and at any point on the walking surface of the deck, and not less than an average of 0.6 foot-candle (0.6 lumen per square foot) [6.46 lux]. A maximum-to-minimum illumination uniformity ratio of 40 to 1 shall not be exceeded.

322.4 Residential pool and deck illumination. Where lighting is installed for, and in, *residential* pools and permanent *residential* spas, such lighting shall be installed in accordance with NFPA 70 or the *International Residential Code*, as applicable in accordance with Section 102.7.1.

SECTION 323—LADDERS AND RECESSED TREADS

323.1 General. Ladders and recessed treads shall comply with the provisions of this section and the applicable provisions of Chapters 4 through 10 based on the type of pool or spa.

323.2 Outside diving envelope. Where installed, steps and ladders shall be located outside of the minimum diving water envelope as indicated in Figure 323.2.

FIGURE 323.2—MINIMUM WATER DIVING ENVELOPE

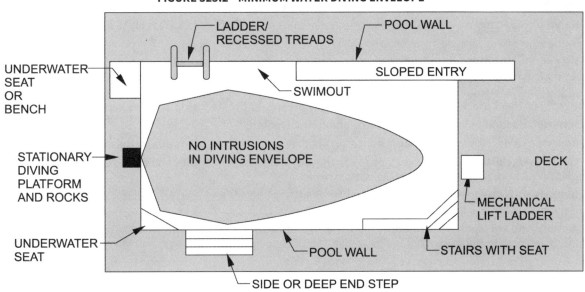

323.3 Ladders. Ladder treads shall have a uniform horizontal depth of not less than 2 inches (51 mm). There shall be a uniform distance between ladder treads, with a distance of not less than 7 inches (178 mm) and not greater than 12 inches (305 mm). The top tread of a ladder shall be located not greater than 12 inches (305 mm) below the top of the deck or coping. Ladder treads shall have slip-resistant surfaces.

323.3.1 Wall clearance. There shall be a clearance of not less than 3 inches (76 mm) and not greater than 4 inches (101.6 mm) between the pool wall and the ladder.

323.3.2 Handrails and handholds. Ladders shall be provided with two handholds or two handrails. The clear distance between ladder handrails shall be not less than 17 inches (432 mm) and not greater than 24 inches (610 mm).

323.4 Recessed treads. Recessed treads shall have a depth of not less than 4.5 inches (114 mm) and a width of not less than 12 inches (305 mm). The vertical distance between the pool coping edge, deck, or step surface and the uppermost recessed tread shall be not greater than 12 inches (305 mm) measured at the wall. The tread shall not protrude more than 2.5 inches (64 mm) from the wall. Recessed treads shall have slip-resistant surfaces.

323.4.1 Vertical spacing. Recessed treads at the centerline shall have a uniform vertical spacing of not less than 7 inches (178 mm) and not greater than 12 inches (305 mm).

323.4.2 Drainage. Recessed treads shall drain into the pool.

323.4.3 Handrails and grab rails. Recessed treads shall be provided with a handrail or grab rail on each side of the treads. The clear distance between handrails and grab rails shall be not less than 17 inches (432 mm) and not greater than 24 inches (610 mm).

SECTION 324—SAFETY

324.1 Handholds required. Where the depth below the *design waterline* of a *residential* swimming pool or spa exceeds 42 inches (1067 mm) or where the depth below the *design waterline* of a public swimming pool exceeds 24 inches (610 mm), handholds along the perimeter shall be provided. Handholds shall be located at the top of deck or coping.

Exceptions:

1. Handholds shall not be required where an underwater bench, seat or swimout is installed.
2. Handholds shall not be required for wave action pools and action rivers.

324.1.1 Height above water. Handholds shall be located not more than 12 inches (305 mm) above the *design waterline*.

324.1.2 Handhold type. Handholds shall be one or more of the following:

1. Top of pool deck or coping.
2. Secured rope.
3. Rail.
4. Rock.
5. Ledge.
6. Ladder.
7. Stair step.
8. Any design that allows holding on with one hand while at the side of the pool.

324.1.3 Handhold spacing. Handholds shall be horizontally spaced not greater than 4 feet (1219 mm) apart.

324.2 Handrails. Where handrails are installed, they shall conform to this section.

324.2.1 Height. The top of the gripping surface of handrails for public pools and public spas shall be 34 inches (864 mm) to 38 inches (965 mm) above the ramp or step surface as measured at the nosing of the step or finished surface of the slope. The top of the gripping surface of handrails for *residential* pools and *residential* spas shall be 30 inches (762 mm) to 38 inches (965 mm) above the ramp or step surface as measured at the nosing of the step or finished surface of the slope.

324.2.2 Material. Handrails shall be made of corrosion-resistant materials.

324.2.3 Nonremovable. Handrails shall be installed so that they cannot be removed without the use of tools.

324.2.4 Leading edge distance. The leading edge of handrails for stairs, pool entries and exits shall be located not greater than 18 inches (457 mm) from the vertical face of the bottom riser.

324.2.5 Diameter. The outside diameter or width of *handrails* shall be not less than $1^1/_4$ inches (32 mm) and not greater than 2 inches (51 mm).

324.3 Obstructions and entrapment avoidance. There shall not be obstructions that can cause the user to be entrapped or injured. Types of entrapment include, but are not limited to, wedge or pinch-type openings and rigid, nongiving cantilevered protrusions.

SECTION 325—EQUIPMENT ROOMS

325.1 General. The provisions of this section apply to public pools and spas and aquatic recreation facilities.

325.2 Requirements. The equipment area or room floor shall be of concrete or other suitable material having a smooth slip-resistant finish and have positive drainage, including a sump drain pump, if necessary. Floors shall have a slope toward the floor drain or sump drain pump adequate to prevent standing water at all times. The opening to the equipment room or area shall be designed to provide access for all anticipated equipment. At least one hose bibb with backflow preventer shall be located in the equipment room or allow for access within an adequate distance of the equipment room so that a hose can service the entire room.

325.3 Construction. The size of the equipment room or area shall provide working space to perform routine operations and equipment service. Equipment rooms also intended for storage shall have adequate space provided for such storage, without reducing the working spaces. Equipment rooms or areas shall be lighted to provide 30 foot-candles (323 lux) of illumination at floor level.

325.4 Electrical. All electrical wiring shall be installed in accordance with NFPA 70.

325.5 Ventilation. Equipment room ventilation shall address all of the following:

1. Combustion requirements.
2. Heat dissipation from equipment.
3. Humidity from surge or balance tanks.
4. Ventilation to the outside.
5. Air quality.

325.6 Markings. All piping in the equipment room shall be permanently identified by its use and the pool or spa it serves. Identification shall be provided for all of the following:

1. Main drains and skimmer lines.
2. Filtered water.
3. Make-up water
4. Chlorine (or disinfection) feeds.
5. Acid (or pH) feeds.
6. Compressed air lines.
7. *Gutter* lines.
8. Chemical sample piping.
9. Pool heating lines.

All piping shall be marked with directional arrows as necessary to determine flow direction and all valves shall be clearly identified by number with a brass tag, plastic laminate tag or permanently affixed alternative. Valves shall be described as to their function and referenced in the operating instruction manual.

325.7 Separation from chemical storage spaces. Combustion equipment, air-handling equipment, and electrical equipment shall not be exposed to air contaminated with corrosive chemical fumes or vapors. Spaces containing combustion equipment, air-handling equipment or electrical equipment and spaces sharing air distribution with spaces containing such equipment shall not be used as chemical storage spaces at the same time unless the equipment is *listed* and *labeled* for use in that atmosphere. Spaces containing combustion equipment, air-handling equipment, or electrical equipment and spaces sharing air distribution with spaces containing such equipment shall be isolated from chemical storage space air.

325.7.1 Doors and openings. A door or doors shall not be installed in a wall between such equipment rooms and an interior chemical storage space. There shall be no ducts, grilles, pass-throughs, or other openings connecting such equipment rooms to chemical storage spaces, except as permitted by the *International Fire Code*.

Spaces containing combustion equipment, air-handling equipment, or electrical equipment and spaces sharing air distribution with spaces containing such equipment shall be isolated from indoor aquatic facility air unless the equipment is listed for the atmosphere. There shall be no ducts, grilles, pass-throughs, or other openings connecting such spaces to an indoor aquatic facility.

Ducts that connect the indoor aquatic facility to the duct connections of air handlers shall not be construed as connecting the air-handler space to the indoor aquatic facility unless HVAC equipment is rated for indoor aquatic facility atmosphere and serves only that indoor aquatic facility.

Where building construction leaves any openings or gaps between floors and walls, or between walls and other walls, or between walls and ceilings, such gaps shall be permanently sealed against air leakage.

325.7.2 Indoor aquatic facility access. Where a door or doors are installed in a wall between an equipment room and an indoor aquatic facility, the floor of the equipment room shall slope back into the equipment room in such a way as to prevent any equipment room spills from running under the door into the indoor aquatic facility. This requirement shall be accomplished by one of the following:

1. A floor all of which is at least 4 inches (102 mm) below the level of the nearest part of the indoor aquatic facility floor.

2.   A continuous dike not less than 4 inches (102 mm) high located entirely within the equipment room, which will prevent spills from reaching the indoor aquatic facility floor.

325.7.2.1 Automatic closer and lock. A door between an equipment room and an indoor aquatic facility shall be equipped with an automatic closer and automatic lock. The door, frame, and automatic closer shall be installed so as to ensure that the door closes completely and latches without human assistance. The automatic lock shall require a key or combination to open from the indoor aquatic facility side. The lock shall be designed and installed to be opened by one hand from the inside of the room under all circumstances, without the use of a key or tool.

Such doors shall be equipped with permanent signage warning against unauthorized entry. All sides of such doors shall be equipped with a gasket. The gasket shall be installed to prevent the passage of air, fumes, or vapors when the door is closed.

325.8 Chemical storage space. A least one space dedicated to chemical storage space shall be provided to allow safe storage of pool and spa chemicals. In all chemical storage spaces, an emergency eyewash station shall be provided. The construction of a chemical storage space shall take into account foreseeable hazards and protect the stored materials against tampering, wildfires, unintended exposure to water and the transfer of fumes into any interior space of a building intended for occupation. Any walls, floors, doors, ceilings, and other building surfaces of an interior chemical storage space shall join each other tightly.

If chemicals are to be stored outdoors, they shall be stored in a well-ventilated protective area with an installed barrier to prevent unauthorized access. Exterior chemical storage spaces not joined to a wall of a building shall be completely enclosed by fencing that is at least 6 feet high (1829 mm). Fencing shall be equipped with a self-closing and self-latching gate having a permanent locking device.

325.8.1 Chemical storage space doors. All doors opening into chemical storage spaces shall be equipped with permanent signage having all of the following:

1.   A warning against unauthorized entry.
2.   Statement of the expected hazards.
3.   Statement of the location of the associated safety data sheet forms.
4.   Product chemical hazard NFPA chart.

Where a single door is the only means of egress from a chemical storage space, the door shall be equipped with an emergency-egress device. Where a chemical storage space door must open to an interior space, spill containment shall be provided to prevent spilled chemicals from leaving the chemical storage space and the door shall not open to a space containing combustion equipment, air-handling equipment, or electrical equipment.

325.8.1.1 Interior opening doors. Where a chemical storage space door must open to an interior space, such door shall have all of the following requirements:

1.   Constructed of corrosion-resistant materials.
2.   Equipped with a corrosion-resistant, automatic lock to prevent unauthorized entry.
    2.1.   Such lock shall require a key or combination to open from the outside into the chemical storage space.
    2.2.   Such lock shall be designed and installed as to be capable of being opened by one hand from the inside of the chemical storage space without the use of a key or tool.
3.   Supported on corrosion-resistant hinges, tracks, or other supports.
4.   Equipped with suitable gaskets or seals on the top and all sides to minimize air leakage between the door and the door frame.
5.   Equipped with a floor or threshold seal to minimize air leakage between the door and the floor or threshold.
6.   Equipped with an automatic door closer that will completely close the door and latch without assistance and against the specified difference in air pressure.
7.   Equipped with a limit switch and an alarm that will sound if the door remains open for more than 30 minutes. The alarm shall have a minimum output level of 85 dbA at 10 feet (2028 mm).

325.8.2 Interior chemical storage spaces. There shall be no transfer grille, pass-through grille, louver, or other device or opening that will allow air movement from the chemical storage space into any other interior space of a building intended for occupancy or into another chemical storage space.

Interior chemical storage spaces that share any building surface with any other interior space shall be equipped with a ventilation system that operates continuously and ensures that all air movement is from all other interior space and toward the chemical storage space.

Interior chemical storage spaces that share an electrical conduit system with any other interior space shall be equipped with a ventilation system that operates continuously and ensures that all air movement is from all other interior spaces and toward the chemical storage space. This pressure difference shall be maintained by a continuously operated exhaust system used for no other purpose than to remove air from that one chemical storage space.

Where more than one chemical storage space is present, a separate exhaust system shall be provided for each chemical storage space. The exhaust airflow rate shall be the amount specified in the *International Mechanical Code*. The function of this exhaust system shall be monitored continuously by an audible differential-pressure alarm system that shall sound if the specified differential air pressure is not maintained for a period of 30 minutes. This alarm shall have a minimum output level of 85 dbA at 10 feet (3048 mm) and shall require manual reset to silence it.

325.8.2.1 Air ducts in interior chemical storage spaces. No duct shall allow air movement from the chemical storage space into any other interior space of a building intended for occupation or into any other chemical storage space.

Air ducts shall not enter or pass through an interior chemical storage space unless it is a corrosion-resistant duct used for no other purpose than to exhaust air from the chemical storage space. This corrosion-resistant duct must exhaust to the exterior and must end at a point on the exterior of the building, at least 20 feet (6096 mm) from any air intake for breathing air, cooling air, or combustion air.

A duct used for no other purpose than to supply makeup air to the chemical storage area shall be acceptable. This makeup air supply duct must end at a point on the exterior of the building, at least 20 feet (6096 mm) from any air intake for breathing air, cooling air, or combustion air.

Any other ducts specifically allowable by the *International Mechanical Code* where such ducts are corrosion resistant and free-of-joints to the extent feasible shall be acceptable.

325.8.2.2 Pipes and tubes in interior chemical storage spaces. Pipes and tubes shall not enter or pass through an interior chemical storage space.

Exceptions:
1. As required to service devices integral to the function of the chemical storage space, such as pumps, vessels, controls, freeze protection, and safety devices.
2. As required to allow for automatic fire suppression.
3. As required for drainage.

Piping, tubes, drain bodies, grates, and attachment and restraint devices shall be corrosion resistant and rated for the chemical environment(s) present, including floor drain bodies and grates. All wall penetrations shall be sealed air tight and commensurate with the rating of the wall assembly. Sealing materials shall be compatible with the wall assembly and the chemical environment(s) present.

325.8.2.3 Combustion equipment in interior chemical storage. No combustion device or appliance shall be installed in a chemical storage space, or in any other place where it will be exposed to the air from a chemical storage space.

Exceptions: A combustion device or appliance that meets all of the following requirements shall be acceptable:
1. The device or appliance is required for one or more processes integral to the function of the room, such as space heat.
2. The device is listed for such use.
3. The device as installed is *approved*.

325.9 Ozone rooms. An ozone equipment room shall not be used for storage of chemicals, solvents, or any combustible materials, other than those required for the operation of the recirculation and ozone-generating equipment. Rooms that are designed to include ozone equipment shall be equipped with an emergency ventilation system capable of 6 air changes per hour. The exhaust intake shall be located 6 inches (152 mm) from the floor, on the opposite side of the room from the make-up air intake.

The emergency ventilation system shall be so arranged as to run on command of an ozone-leak alarm or on command of a manual switch. The manual emergency ventilation switch shall be located outside the room and near the door to the ozone room.

Ozone rooms which are below grade shall be equipped with force-draft ventilation capable of 6 changes per hour. The exhaust intake shall be located 6 inches (152 mm) from the floor, on the opposite side of the room from the make-up air intake. Such ventilation shall be so arranged as to:
1. Run automatically concurrent with the ozone equipment and for at least a time allowing for 15 air changes after the ozone equipment is stopped.
2. Run on activation of the ozone detection and alarm system.
3. Run on command of a manual switch.

The manual ventilation switch shall be located outside the room and near the door to the ozone room.

325.9.1 Signage. In addition to the signs on all chemical storage areas, a sign shall be posted on the exterior of the entry door stating "DANGER - GASEOUS OXIDIZER -- OZONE" in lettering not less than 4 inches (102 mm) high.

325.9.2 Alarm system. Rooms containing ozone-generation equipment shall be equipped with an audible and visible ozone detection and alarm system.

The alarm system shall consist of both an audible alarm capable of producing at least 85 decibels at a 10-foot (3048 mm) distance and a visible alarm consisting of a flashing light mounted in plain view of the entrance to the ozone equipment room.

The ozone sensor shall be located at a height of 18-24 inches (457-610 mm) above floor level. The ozone sensor shall be capable of measuring ozone in the range of 0-2 ppm.

The alarm system shall activate when the ozone concentration equals or exceeds 0.1 ppm in the room. Activation of the alarm system shall shut off the ozone-generating equipment and turn on the emergency ventilation system.

325.10 Gaseous chlorination space. Use of compressed chlorine gas shall be prohibited for new construction and after substantial *alteration* to existing facilities.

325.11 Windows. Where windows are installed in an interior wall, ceiling, or door of a chemical storage space, such windows shall have the following components:

1. Tempered or plasticized glass.
2. A corrosion-resistant frame.
3. Incapable of being opened or operated.

Where windows are installed in an exterior wall or ceiling, such windows shall:

1. Be mounted in a corrosion-resistant frame.
2. Be protected by a roof, eave, or permanent awning as to minimize the entry of rain or snow in the event of window breakage.

325.12 Sealing and blocking materials. Materials used for sealing and blocking openings in an interior chemical storage space shall:

1. Minimize the leakage of air, vapors, or fumes from the chemical storage space.
2. Be compatible for use in the environment.
3. Be commensurate with the fire rating assembly in which they are installed.

SECTION 326—INDOOR AIR QUALITY

326.1 General. Indoor public pool and spa air-handling system design, construction, and installation shall comply with the requirements of the *International Mechanical Code* or ASHRAE 62.1.

PUBLIC SWIMMING POOLS

User notes:

About this chapter: *Chapter 4 has regulations for public swimming pools. Where diving boards are present, the chapter provides information regarding the minimum diving water dimensions. Requirements for exiting pools, decks, circulation systems and depth markers are provided. Special features of pools such as rest ledges, swimouts and underwater seats and benches are regulated by this chapter.*

SECTION 401—GENERAL

401.1 Scope. The provisions of this chapter shall apply only to Class A, Class B, Class C, Class E and Class F public swimming pools.

401.2 Intent. The provisions in this chapter shall govern the design, equipment, operation, warning signs, installation, sanitation, new construction, and *alteration* specific to the types of public swimming pools indicated in Section 401.1.

401.3 Chapter 3 compliance required. In addition to the requirements of this chapter, public swimming pools shall comply with the requirements of Chapter 3.

401.4 Dimensional tolerances. Finished pool dimensions, for other than Class A pools, shall be held within the construction tolerances shown in Table 401.4. Other dimensions, unless otherwise specified, shall have a tolerance of ± 2 inches (51 mm).

TABLE 401.4—CONSTRUCTION TOLERANCES	
DESIGN ASPECT	**CONSTRUCTION TOLERANCE**
Depth–deep area, including diving area	± 3 inches
Depth–shallow area	± 2 inches
Length–overall	± 3 inches
Step treads & risers	± $\frac{1}{2}$ inch
Wall slopes	± 3 degrees
Waterline–pools with adjustable weir skimmers	± $\frac{1}{4}$ inch
Waterline–pools with nonadjustable skimming systems (gutters)	± $\frac{1}{8}$ inch
Width–overall	± 3 inches
All dimensions not otherwise specified herein	± 2 inches
For SI: 1 inch = 25.4 mm, 1 degree = 0.017 radians.	

401.4.1 Class A pool tolerances. Dimensional tolerances for Class A pools shall be determined by the authority that provides the accreditation of the pool for competitive events.

401.5 Floor slope. Except where required to meet the accessibility requirements in accordance with Section 307.9, the slope of the floor in the shallow area of a pool shall not exceed 1 unit vertical in 10 units horizontal (10-percent slope) for Class C pools and 1 unit vertical in 12 units horizontal (8-percent slope) for Class B pools. The slope limit shall apply in any direction to the point of the first slope change, where a slope change exists. The point of the first slope change shall be defined as the point at which the floor slope exceeds 1 unit vertical in 10 units horizontal (10-percent slope) for Class C pools and 1 unit vertical in 12 units horizontal (8-percent slope) for Class B pools.

401.6 Dimensions for Class A pools. Class A pools shall be designed and constructed with the dimensions determined by the authority that provides the accreditation of the pool for competitive events.

SECTION 402—DIVING

402.1 General. This section covers diving requirements for Class B, Class C, and Class E pools. Manufactured and fabricated diving equipment and appurtenances shall not be installed on Type O pools.

402.2 Manufactured and fabricated diving equipment. Manufactured and fabricated diving equipment shall be in accordance with this section and shall be designed for swimming pool use.

402.3 Installation. The installation of manufactured diving equipment shall be in accordance with Sections 402.3 through 402.14. Manufactured diving equipment shall be located in the deep area of the pool so as to provide the minimum dimensions shown in Table 402.12 and shall be installed in accordance with the manufacturer's instructions. Installation and use instructions for manufactured diving equipment shall be provided by the manufacturer and shall specify the minimum diving water envelope dimensions required for each diving board and diving stand combination. The manufacturer's instructions shall refer to the water envelope type by dimensionally relating their products to Point A on the diving water envelopes shown in Table 402.12. The diving board manufacturer shall specify which boards fit on the design pool geometry types as indicated in Table 402.12.

402.4 Slip resistance. Diving equipment shall have slip-resistant walking surfaces.

402.5 Point A. For the application of Table 402.12, Point A shall be the point from which dimensions of width, length and depth are established for the minimum diving water envelope. If the tip of the diving board or diving platform is located at a distance of WA (see Figure 804.1) or greater from the deep end wall and the water depth at that location is equal to or greater than the water depth requirement at Point A, the point on the water surface directly below the center of the tip of the diving board or diving platform shall be identified as Point A.

402.6 Location of pool features in a diving pool. Where a pool is designed for use with diving equipment, the location of steps, pool stairs, ladders, underwater benches, underwater ledges, special features and other accessory items shall be outside of the minimum diving water envelope. See Figure 323.2.

402.7 Stationary diving platforms and diving rocks. Where stationary diving platforms and diving rocks are built on site, flush with the wall and located in the diving area of the pool, Point A shall be in front of the wall at the platform or diving rock centerline.

402.8 Location of diving equipment. Manufactured and fabricated diving equipment shall be located so that the tip of the board or platform is located directly above Point A as defined by Section 402.5.

402.9 Elevation. The maximum elevation of a diving board above the *design waterline* shall be in accordance with the manufacturer's instructions.

402.10 Platform height above waterline. The height of an *approved* stationary diving apparatus, platform, or diving rock above the *design waterline* shall not exceed the limits of the manufacturer's specifications or the limits of the design prepared by a design professional.

402.11 Clearance. The diving equipment manufacturer shall specify the minimum headroom required above the tip of the board.

402.12 Water envelopes. The minimum diving water envelopes shall be in accordance with Table 402.12.

TABLE 402.12—MINUMUM DIVING WATER ENVELOPES (See Figure 402.12)											
POOL TYPE	**MINIMUM DIMENSIONS**								**MINIMUM WIDTH OF POOL AT:**		
	D_1	D_2	R	L_1	L_2	L_3	L_4	L_5	Pt. A	Pt. B	Pt. C
VI	7'-0"	8'-6"	5'-6"	2'-6"	8'-0"	10'-6"	7'-0"	28'-0"	16'-0"	18'-0"	18'-0"
VII	7'-6"	9'-0"	6'-0"	3'-0"	9'-0"	12'-0"	4'-0"	28'-0"	18'-0"	20'-0"	20'-0"
VIII	8'-6"	10'-0"	7'-0"	4'-0"	10'-0"	15'-0"	2'-0"	31'-0"	20'-0"	22'-0"	22'-0"
IX	11'-0"	12'-0"	8'-6"	6'-0"	10'-6"	21'-0"	0	37'-6"	22'-0"	24'-0"	24'-0"
For SI: 1 inch = 25.4 mm, 1 foot = 304.8 mm.											

FIGURE 402.12—(MINIMUM DIVING WATER ENVELOPES)
CONSTRUCTION DIMENSIONS FOR WATER ENVELOPES FOR CLASS B AND CLASS C POOLS

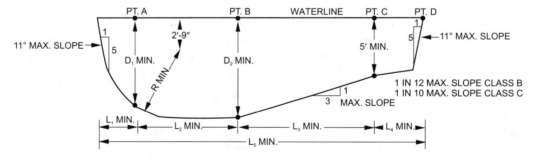

402.13 Ladders for diving equipment. Ladders shall be provided with two grab rails or two handrails. There shall be a uniform distance between ladder treads, with a 7-inch (178 mm) minimum distance and a 12-inch (305 mm) maximum distance.

> **Exception:** The distance between treads for the top and bottom riser can vary but shall be not less than 7 inches (178 mm) and not greater than 12 inches (305 mm).

402.14 Springboard fall protection guards. Springboards located at a height greater than 5 feet (1524 mm) above the pool deck shall have a fall protection guard on each side of the springboard. The design and the selection of the materials of construction of the fall protection guards shall be determined by the manufacturer of the springboard support structure. The installation and maintenance of the fall protection guards shall be in accordance with the fall protection guard manufacturer's instructions.

SECTION 403—BATHER LOAD

403.1 Maximum bather load. The maximum bather load of Class B and Class C pools shall be in accordance with Table 403.1.

TABLE 403.1—MAXIMUM BATHER LOAD			
POOL/DECK AREA	**SHALLOW INSTRUCTIONAL OR WADING AREAS**	**DEEP AREA (NOT INCLUDING THE DIVING AREA)**	**DIVING AREA (PER EACH DIVING BOARD)**
Pools with minimum deck area	15 sq. ft. per user	20 sq. ft. per user	300 sq. ft.
Pools with deck area at least equal to water surface area	12 sq. ft. per user	15 sq. ft. per user	300 sq. ft.
Pools with deck area at least twice the water surface area	8 sq. ft. per user	10 sq. ft. per user	300 sq. ft.
For SI: 1 square foot = 0.09 square meters.			

SECTION 404—REST LEDGES

404.1 Rest ledges. Rest ledges along the pool walls are permitted. They shall be not less than 4 feet (1220 mm) below the water surface. Where a ledge is provided, the width of the ledge shall be not less than 4 inches (102 mm) and not greater than 6 inches (152 mm).

SECTION 405—WADING POOLS

405.1 Wading pools. Class F wading pools shall be separate pools with an independent circulation system, shall be physically separated from the main pool and shall be constructed in accordance with Sections 405.2 through 405.6.

405.2 Nonentry areas. The areas where the water depth at the edge of the pool exceeds 9 inches (229 mm) shall be considered to be nonentry areas.

405.3 Floor slope. The floors of wading pools shall be uniform and sloped with a maximum slope of 1 unit vertical in 12 units horizontal (8-percent slope).

405.4 Maximum depth. The water depth shall not exceed 18 inches (457 mm).

405.5 Distance from deck to waterline. The maximum distance from the top of the deck to the waterline shall not exceed 6 inches (152 mm).

405.6 Suction entrapment avoidance. Suction outlet fitting assemblies shall not be located in wading pools where bathers have access to such outlets. Where suction outlets cannot be located to avoid bather access, skimmers or overflow *gutters* shall be installed and shall accommodate 100 percent of the circulation system flow rate.

SECTION 406—DECKS AND DECK EQUIPMENT

406.1 General. Decks shall comply with the provisions of Section 306, except as otherwise required in this section.

406.2 Pool perimeter access. A deck or unobstructed access shall be provided for not less than 90 percent of the pool perimeter.

406.3 Deck clearance. Decking not less than 4 feet (1219 mm) in width shall be provided on the sides and rear of any diving equipment. A deck clearance of 4 feet (1219 mm) shall be provided around all other deck equipment.

406.4 Decks between pools and spas. Decks between pools, spas or any combination of pools and spas, shall have a width of not less than 6 feet (1829 mm).

406.5 Deck covering. Walking surfaces of decks within 4 feet (1219 mm) of a pool or spa shall be slip resistant.

406.6 Distances above diving boards. A completely unobstructed minimum distance above the tip of the diving board shall be specified by the diving equipment manufacturer.

406.7 Dimensional requirements. Public pools with diving equipment of 39 inches (991 mm) or greater in height, and pools designed for springboard or platform diving, shall comply with the dimensional design requirements of the diving equipment manufacturer or the authority that governs such pools.

406.8 Diving equipment. Diving equipment shall be installed in accordance with the manufacturer's specifications.

    406.8.1 Label. A label shall be permanently affixed to the diving equipment or jump board in a readily visible location and shall include all of the following:

        1.   The minimum diving water envelope required for each diving board and diving stand combination.

        2.   Manufacturer's name and address.

        3.   Manufacturer's identification and date of manufacture.

        4.   The maximum allowable weight of the user.

    406.8.2 Use instructions. The diving equipment manufacturer shall provide diving equipment use instructions.

406.8.3 Tread surface. Diving equipment shall have slip-resistant tread surfaces.

406.8.4 Supports for diving equipment. Supports, platforms, stairs, and ladders for diving equipment shall be designed to carry the anticipated loads. Stairs and ladders shall be of corrosion-resistant materials, shall be easily cleanable and shall have slip-resistant treads. Diving stands higher than 21 inches (533 mm), measured from the deck to the top back end of the board, shall be provided with stairs or a ladder. Step treads shall be self-draining.

406.8.5 Guardrails. Diving equipment 39 inches (991 mm) or greater in height shall be provided with a top guardrail. Such guardrail shall extend not less than 30 inches (762 mm) above the diving board and extend to the edge of the pool wall.

406.9 Starting blocks. In new construction or substantial *alteration*, starting blocks intended for competitive swimming shall be located at a water depth of not less than 5 feet (1524 mm).

406.10 Swimming pool slides. Swimming pool slides shall comply with the requirements of 16 CFR, Part 1207. The manufacturer of the slide shall provide installation and use instructions for the slide. Slides shall be installed in accordance with the manufacturer's instructions.

406.11 Play and water activity equipment. Play and water activity equipment shall be installed in accordance with the manufacturer's instructions.

SECTION 407—CIRCULATION SYSTEMS

407.1 General. Circulation systems for pools shall comply with Section 312 and the provisions of this section.

407.2 Turnover. Circulation equipment shall be sized to turn over the entire water capacity of the pool as specified in Table 407.2. The system shall be designed to provide the required turnover rate based on the maximum pressure and flow rate recommended by the manufacturer of the filter with clean filter media.

TABLE 407.2—TURNOVER RATE	
SWIMMING POOL CATEGORY	**TURNOVER RATE IN HOURS**
Class A, B, and C pools	Hours equal $1^1/_2$ times the average depth of pool in feet not to exceed 6 hours
Wading pools	1
For SI: 1 foot = 304.8 mm.	

407.3 Continuous water removal. The design of a *gutter* system shall accommodate continuous removal of water from the pool's upper surface at a rate of not less than 125 percent of the required total recirculation flow rate as determined by the design professional.

407.3.1 Gutter outlets. At a *gutter* flow condition of not less than 125 percent of the total recirculation flow as determined by the design professional, *gutter* outlets such as drop boxes, converters, return piping, or flumes shall be designed to prevent flooding of the *gutter* that would result in skimmed water reentering the pool.

407.3.2 Adequate mixing. Pools shall have wall or floor inlets or both to provide for adequate mixing. Inlets shall be hydraulically sized to provide the design flow rates for each area of the pool proportional to the turnover rate and the area covered by the inlet.

407.4 Pool circulation. The filtration circulation system shall be designed with sufficient flexibility to achieve a hydraulic apportionment that will ensure effective distribution of treated water throughout the pool.

407.4.1 Inlets. Effective distribution of treated water shall be accomplished by either a continuous perimeter overflow system with integral inlets or by means of directionally adjustable inlets adequate in design, number, and location.

407.4.2 Adequate mixing. Pools shall have wall or floor inlets or both to provide for adequate mixing. Inlets shall be hydraulically sized to provide the design flow rates for each area of the pool proportional to the turnover rate and the area covered by the inlet.

407.4.3 Floor inlets. Floor inlets shall be required for pools that are greater than 50 feet (15.2 m) wide. The spacing between floor inlets shall not exceed 20 feet (6.1 m). Pools having only floor inlets shall have such inlets located within 15 feet (4.6 m) of the perimeter waterline. Where wall inlets are used in combination with floor inlets, the floor inlets shall be located not greater than 25 feet (7.6 m) from the nearest side walls.

407.4.4 Wall inlets. The spacing between wall inlets shall not exceed 20 feet (6.1 m), measured along the perimeter waterline.

SECTION 408—FILTERS

408.1 General. Filters shall be designed in accordance with Section 313, except as otherwise required in this section.

408.2 Air release warning. The following statement shall be posted in a conspicuous location within the areas of the air release:

DO NOT START THE SYSTEM AFTER MAINTENANCE
WITHOUT FIRST PROPERLY REASSEMBLING THE FILTER AND
SEPARATION TANK AND OPENING ALL AIR RELEASE VALVES.

SECTION 409—SPECIFIC SAFETY FEATURES

409.1 Handholds. Handholds shall comply with the provisions of Section 324.

409.2 Depth markers. Depth markers shall be provided in accordance with Sections 409.2.1 through 409.2.8.

409.2.1 Where required. Depth markers shall be installed at the maximum and minimum water depths and at all points of slope change. Depth markers shall be installed at water depth increments not to exceed 2 feet (607 mm). Depth markers shall be spaced at intervals not to exceed 25 feet (7620 mm).

409.2.2 Marking of depth. The depth of water in feet (meters) shall be plainly and conspicuously marked on the vertical pool wall at or above the waterline.

Exception: Pools with a vanishing edge and rim flow *gutters*.

409.2.3 Depth accuracy. Depth markers shall indicate the actual pool depth within ± 3 inches (76 mm), at normal operating water level where measured 3 feet (914 mm) from the pool wall or at the tangent point where the cove radius meets the floor, whichever is deeper.

409.2.4 Position on pool wall. Depth markers on the vertical pool wall shall be positioned to be read from the waterside. Depth markers shall be placed so as to allow as much of the numbers to be visible above the waterline as possible.

409.2.5 Position on deck. Depth markers on the deck shall be located within 18 inches (457 mm) of the water edge and positioned to be read while standing on the deck facing the water.

409.2.6 Horizontal markers. Horizontal depth markers shall be slip resistant.

409.2.7 Uniform distribution. Depth markers shall be distributed uniformly on both sides and both ends of the pool.

409.2.8 Numbers and letters. Depth markers shall be not less than 4 inches (102 mm) in height. The color of the numbers shall contrast with the background on which they are applied and the color shall be of a permanent nature. The lettering shall spell out the words "feet" and "inches" or abbreviate them as "Ft." and "In." respectively. Where displayed in meters in addition to feet and inches, the word meter shall be spelled out or abbreviated as "M."

409.3 No diving symbol. Where the pool depth is 5 feet (1524 mm) or less, the "No Diving" symbol shall be displayed. The symbol shall be placed on the deck at intervals of not greater than 25 feet (7620 mm) and directly adjacent to a depth marker. Additional signage shall be in accordance with NEMA Z535.

409.4 Lifesaving equipment. Public pool Classes A, B, and C shall be provided with lifesaving equipment in accordance with Sections 409.4.1 through 409.4.3. Such lifesaving equipment shall be visually conspicuous and conveniently located at all times.

409.4.1 Accessory pole. A swimming pool accessory pole not less than 12 feet (3658 mm) in length and including a body hook shall be provided.

409.4.2 Throwing rope. A throwing rope attached to a ring buoy or similar flotation device shall be provided. The rope shall be not less than $^1/_4$ inch (6.4 mm) in diameter and shall have a length of not less than $1^1/_2$ times the maximum width of the pool or 50 feet (15 240 mm), whichever is less. A ring buoy shall have an outside diameter of not less than 15 inches (381 mm).

409.4.3 Emergency response units. Pools covered by this chapter shall be provided with first aid equipment, including a first aid kit. First aid equipment and kits shall be located to allow access.

SECTION 410—SANITARY FACILITIES

410.1 Toilet facilities. Class A and B pools shall be provided with toilet facilities having the required number of plumbing fixtures in accordance with the *International Building Code* or the *International Plumbing Code*.

SECTION 411—SPECIAL FEATURES

411.1 Entry and exit. Pools shall have not less than two means of entry and exit that are located so as to serve both ends of a pool. Pool lifts, transfer walls and transfer systems that provide for pool entry and exit by persons with physical disabilities in accordance with Section 307.1.5 shall not be counted as the means of entry or exit that is required by this section.

411.1.1 Natural entry. Where areas have water depths of 24 inches (607 mm) or less at the pool wall, such areas shall be considered to be providing their own natural mode for entry and exit.

Exception: Wading pools as outlined in Section 405.

411.1.2 Shallow area. A means of entry and exit shall be provided in shallow areas of pools and shall consist of pool stairs, a ramp or a beach entry.

411.1.3 Deep area. The means of entry and exit in the deep area of pools shall consist of one of the following:

1. Steps/stairs.
2. Ladders.
3. Grab rails with recessed treads.
4. Ramps.
5. Beach entries.
6. Swimouts.
7. Other designs that provide the minimum utility as specified in this code.

411.1.4 Pools greater than 30 feet wide. Swimming pools greater than 30 feet (9144 mm) in width shall be provided with entries and exits on each side of the deep area of the pool. The entries and exits on the sides of the deep area of a pool shall be located not more than 82 feet (25 m) apart.

411.1.5 Diving envelope. Where the pool is designed for use with diving equipment, the entries and exits, pool stairs, ladders, underwater benches, special features and other accessories shall be located outside of the minimum diving water envelope indicated in Figure 323.2.

411.1.6 Treads. Treads shall have slip-resistant surfaces.

411.2 Pool stairs. The design and construction of stairs extending into the pool in either shallow or deep water, including recessed pool stairs, shall comply with Sections 411.2.1 through 411.2.4.

411.2.1 Tread dimensions and area. Treads shall be not less than 24 inches (607 mm) at the leading edge. Treads shall have an unobstructed surface area of not less than 240 square inches (0.154 m²) and an unobstructed horizontal depth of not less than 10 inches (254 mm) at the centerline.

411.2.2 Risers. Risers, except for the bottom riser, shall have a uniform height of not greater than 12 inches (305 mm) measured at the centerline. The bottom riser height is allowed to vary to the floor.

411.2.3 Top tread. The vertical distance from the pool coping, deck, or step surface to the uppermost tread shall be not greater than 12 inches (305 mm).

411.2.4 Bottom tread. Where stairs are located in water depths greater than 48 inches (1219 mm), the lowest tread shall be not less than 48 inches (1219 mm) below the deck and shall be recessed in the pool wall.

411.2.5 Outlined edges. The leading horizontal and vertical edges of stair treads shall be outlined with slip-resistant contrasting tile or other permanent marking of not less than 1 inch (25.4 mm) and not greater than 2 inches (50.8 mm).

411.3 Shallow end detail for beach and sloping entries. Sloping entries used as a pool entrance shall have a maximum slope of 1 unit vertical in 10 units horizontal (10-percent slope).

411.3.1 Benches and steps. Where benches are used in conjunction with sloping entries, the vertical riser distance shall not exceed 12 inches (305 mm). Where steps are used in conjunction with sloping entries, the requirements of Section 411.2 shall apply.

411.3.2 Vertical drops. A vertical drop exceeding 12 inches (305 mm) within a sloping entry shall be provided with a handrail.

411.3.3 Surfaces. Beach and sloping entry walking surfaces at water depths up to 36 inches (914 mm) shall be slip resistant.

411.4 Pool ladder design and construction. The design and construction of ladders shall comply with Section 323.

411.5 Underwater seats, benches, and swimouts. The design and construction of underwater seats, benches, and swimouts shall comply with Sections 411.5.1 and 411.5.2.

411.5.1 Swimouts. Swimouts, located in either the deep or shallow area of a pool, shall comply with all of the following:

1. The horizontal surface shall be not greater than 20 inches (508 mm) below the waterline.
2. An unobstructed surface shall be provided that is equal to or greater than that required for the top tread of the pool stairs in accordance with Section 411.2.
3. Where used as an entry and exit access, swimouts shall be provided with steps that comply with the pool stair requirements of Section 411.2.
4. The leading edge shall be visibly set apart.

411.5.2 Underwater seats and benches. Underwater seats and benches, whether used alone or in conjunction with pool stairs, shall comply with all of the following:

1. The horizontal surface shall be not greater than 20 inches (508 mm) below the waterline.
2. An unobstructed surface shall be provided that is not less than 10 inches (254 mm) in depth and not less than 24 inches (607 mm) in width.
3. Underwater seats and benches shall not be used as the required entry and exit access.
4. Where underwater seats are located in the deep area of the pool where manufactured or constructed diving equipment is installed, such seats shall be located outside of the minimum diving water envelope for diving equipment.
5. The leading edge shall be visually set apart.
6. The horizontal surface shall be at or below the waterline.
7. A tanning ledge or sun shelf used as the required entry and exit access shall be located not greater than 12 inches (305 mm) below the waterline.

SECTION 412—SIGNAGE

412.1 Safety signage. Safety signage advising on the danger of diving into *shallow areas* and on the prevention of drowning shall be provided as required by the authority that governs such pools. Safety signage shall be as shown in Figure 412.1 or similar thereto.

FIGURE 412.1—SAFETY SIGN

412.2 Emergency telephone signs. A sign indicating the location of the nearest landline telephone that can be used to call emergency services shall be posted within sight of the main entry into a pool facility. The sign shall indicate the telephone numbers, including area code, that can be called for emergency services including, but not limited to, police, fire, ambulance and rescue services. If "9-1-1" telephone service is available for any of those services, "9-1-1" shall be indicated next to the telephone number for such services. The sign shall include the street address and city where the pool is located. The nearest landline telephone indicated by the sign shall be one that can be used free of charge to call for emergency services. A sign with the telephone number and address information required by this section shall be posted within sight of the landline telephone.

412.3 Sign placement. Signs shall be positioned for effective visual observation by users as required by the authority that governs such pools.

412.4 Emergency shutoff switch. Signs shall be posted that clearly indicate the location of the pump emergency shutoff switch. Such switch shall be clearly identified as the pump emergency shutoff switch.

CHAPTER 5

PUBLIC SPAS AND PUBLIC EXERCISE SPAS

User notes:

About this chapter: Chapter 5 regulates the depth, seating depth and floor slope of public pools and public exercise spas. Suction fitting, heater and depth marker requirements are also included in this chapter.

SECTION 501—GENERAL

501.1 Scope. This chapter shall govern the design, installation, construction and repair of public spas and exercise spas regardless of whether a fee is charged for use.

501.2 General. In addition to the requirements of this chapter, public spas and public exercise spas shall comply with the requirements of Chapter 3.

SECTION 502—MATERIALS

502.1 Pumps and motors. Pumps and motors shall be *listed* and *labeled* for use in spas.

SECTION 503—STRUCTURE AND DESIGN

503.1 Water depth. The maximum water depth for spas shall be 4 feet (1219 mm) measured from the *design waterline* except for spas that are designed for special purposes and *approved* by the authority having jurisdiction. The water depth for exercise spas shall not exceed 6 feet 6 inches (1981 mm) measured from the *design waterline*.

503.2 Multilevel seating. Where multilevel seating is provided, the maximum water depth of any seat or sitting bench shall be 28 inches (711 mm) measured from the *design waterline* to the lowest measurable point.

503.3 Floor slope. The slope of the floor shall not exceed 1 unit vertical in 12 units horizontal (8.3-percent slope). Where multilevel floors are provided, the change in depth shall be indicated.

SECTION 504—PUMPS AND MOTORS

504.1 Emergency shutoff switch. One emergency shutoff switch shall be provided to disconnect power to circulation and jet system pumps and air blowers. Emergency shutoff switches shall be provided with access. Such switches shall be located within sight of the spa and shall be located not less than 5 feet (1524 mm) but not greater than 10 feet (3048 mm) horizontally from the inside walls of the spa.

504.1.1 Alarms. Emergency shutoff switches shall be provided with an audible alarm rated at not less than 80 decibel sound pressure level and a light near the spa that will operate continuously until deactivated when the shutoff switch is operated. The following statements shall appear on a sign that is posted in a location that is visible from the spa:

ALARM INDICATES SPA PUMPS OFF.
DO NOT USE SPA WHEN ALARM SOUNDS AND
LIGHT IS ILLUMINATED UNTIL ADVISED OTHERWISE.

504.2 Timer. The operation of the hydrotherapy jets shall be limited by a cycle timer having a maximum setting of 10 minutes. The cycle timer shall be located not less than 5 feet (1524 mm) away, adjacent to, and within sight of the spa.

SECTION 505—RETURN AND SUCTION FITTINGS

505.1 Return fittings. Return fittings shall be provided and arranged to facilitate a uniform circulation of water and maintain a uniform sanitizer residual throughout the entire spa or exercise spa.

505.2 Suction fittings. Suction fittings shall be in accordance with Sections 505.2.1 through 505.2.4.

505.2.1 Required conformance. Suction outlet fittings shall be in accordance with Section 312.4.4.

505.2.2 Installation. Suction fittings shall be sized and installed in accordance with the manufacturer's specifications. Spas and exercise spas shall not be used or operated if the suction outlet cover is missing, damaged, broken or loose.

505.2.3 Outlets per pump. Suction fittings shall be provided in accordance with Section 311.

505.2.4 Submerged vacuum fittings. *Submerged vacuum fittings* shall be in accordance with Section 311.

SECTION 506—HEATER AND TEMPERATURE REQUIREMENTS

506.1 General. This section pertains to fuel-fired and electric appliances used for heating spa or exercise spa water.

506.2 Water temperature controls. Components provided for water temperature controls shall be suitable for the intended application.

2024 INTERNATIONAL SWIMMING POOL AND SPA CODE®

506.2.1 Water temperature regulating controls. Water temperature regulating controls shall comply with UL 873 or UL 372. A means shall be provided to indicate the water temperature in the spa.

> **Exception:** Water temperature regulating controls that are integral to the heating appliance and *listed* in accordance with the applicable end use appliance standard.

506.2.2 Water temperature limiting controls. Water temperature limiting controls shall comply with UL 873 or UL 372. Water temperature at the heater return outlet shall not exceed 140°F (60°C).

SECTION 507—WATER SUPPLY

507.1 Water temperature. The temperature of the incoming makeup water shall not exceed 104°F (40°C).

SECTION 508—SAFETY FEATURES

508.1 Instructions and safety signs. Instructions and safety signage shall comply with the requirements of the local jurisdiction. In the absence of local requirements, safety signs and instructions shall comply with UL 1563 or CSA C22.2 No. 218.1.

508.2 Operational signs. Operational signs shall include, but not be limited to, the following messages as required by the local jurisdiction:

1. Do not allow the use of or operate spa if the suction outlet cover is missing, damaged or loose.
2. Check spa temperature before each use. Do not enter the spa if the temperature is above 104°F (40°C).
3. Keep breakable objects out of the spa area.
4. Spa shall not be operated during severe weather conditions.
5. Never place electrical appliances within 5 feet (1524 mm) of the spa.
6. No diving.

508.3 Depth markers. Public spas shall have permanent depth markers with numbers not less than 4 inches (102 mm) in height that are plainly and conspicuously visible from obvious points of entry and in conformance to this section.

508.3.1 Number. There shall be not less than two depth markers for each spa, regardless of spa size or shape.

508.3.2 Spacing. Depth markers shall be spaced at not more than 25-foot (7620 mm) intervals and shall be uniformly located around the perimeter of the spa.

508.3.3 Marking. Spas and exercise spas shall have the maximum water depth clearly marked on the required surfaces and such markers shall be positioned on the deck within 18 inches (457 mm) of the *design waterline*. Depth markers shall be positioned to be read while standing on the deck facing the water.

508.3.4 Slip resistant. Depth markers in or on the deck surfaces shall be slip resistant.

508.4 Clock. Public facilities shall have a clock that is visible to spa users.

AQUATIC RECREATION FACILITIES

User notes:

About this chapter: *Chapter 6 covers facilities commonly known as water parks. Such facilities can have a variety of pools ranging from zero depth entry wave pools to lazy rivers. This chapter includes requirements for floor slopes, steps, marking, signage, circulation systems, handholds and swimouts.*

SECTION 601—GENERAL

601.1 Scope. This chapter covers public pools and water containment systems used for aquatic recreation. This chapter provides specifications for the design, equipment, operation, signs, installation, sanitation, new construction, and rehabilitation of public pools for aquatic play. This chapter covers Class D-1 through Class D-6 public pools whether they are provided as stand-alone attractions or in various combinations in a composite attraction.

601.2 Combinations. Where combinations of Class D-1 through Class D-6 pools exist within a facility, each element in the facility shall comply with the applicable code sections as if the element functioned as a part of a freestanding pool of Class D-1 through Class D-6.

601.3 General. In addition to the requirements of this chapter, aquatic recreation facilities shall comply with the requirements of Chapter 3.

SECTION 602—FLOORS

602.1 Floor slope. In water depths of less than 5 feet (1524 mm), the floor slope shall be not greater than 1 unit vertical in 12 units horizontal (8.3-percent slope) except where the function of the attraction requires greater slopes in limited areas.

> **Exception:** The slope of the floor in Class D-3 pools shall not exceed 1 unit vertical in 7 units horizontal (14-percent slope).

SECTION 603—MARKINGS AND INDICATORS

603.1 Markings. Markings in areas of deep water shall comply with Section 409.2 except where the function of the pool dictates otherwise.

603.2 Class D-2 pools. Where a Class D-2 pool has a bather depth greater than $4^1/_2$ feet (1372 mm), the floor shall have a distinctive marking at the $4^1/_2$ feet (1372 mm) water depth.

603.3 Shallow-to-deep-end rope and float line. Where a pool has a water depth ranging from less than 5 feet (1524 mm) to greater than 5 feet (1524 mm), a rope and float line shall be located 1 foot (305 mm) horizontally from the 5-foot (1524 mm) depth location, toward the shallow end of the pool.

603.4 Nozzles. Pools having nonflush propulsion nozzles in the floor shall have a distinctive marking at the location of such nozzles.

SECTION 604—CIRCULATION SYSTEMS

604.1 General. A circulation system consisting of pumps, piping, return inlets and suction outlets, filters, and other necessary equipment shall be provided for complete circulation of water with the pool.

604.2 Turnover. Circulation system equipment shall be designed to turn over 100 percent of the nominal pool water volume in the amount of time specified in Table 604.2. The system shall be designed to give the required turnover time based on the manufacturer's recommended maximum pressure and flow of the filter in clean media condition.

TABLE 604.2—TURNOVER TIME	
CLASS OF POOL	**MAXIMUM TURNOVER TIME[a] (hours)**
D-1	2
D-2 with less than 24 inches water depth	1
D-2 with 24 inches or greater water depth	2
D-3	1
D-4	2
D-5	1
D-6	1

For SI: 1 inch = 25.4 mm.
a. Pools with a sand bottom require a 1-hour turnover time.

604.2.1 24-hour circulation required. Circulation systems shall circulate treated and filtered water for 24 hours a day.

604.2.2 Reduced circulation rate. The circulation rate shall be permitted to be reduced during periods that the pool is closed for use provided that acceptable water clarity conditions are met prior to reopening the pool for public use. The reduced circulation rate shall not be zero.

604.3 Surface skimming systems. Surface skimming systems shall be in accordance with Table 604.3.

TABLE 604.3—SURFACE SKIMMING SYSTEMS	
CLASS OF POOL	**SURFACE SKIMMING SYSTEM**
D-1	Zero-depth trench located at static water level or other skimming systems
D-2	Auto skimmer, zero-depth trench or gutters
D-3	Auto skimmer, zero-depth trench or perimeter device
D-4	Single or multiple skimmer devices for skimming flow
D-5	Skimmers prohibited in side area
D-6	Auto skimmer, zero-depth trench, or gutters

604.3.1 Class D-5 pool skimmers. The installation of skimmers in the side areas of Class D-5 pools shall be prohibited.

SECTION 605—HANDHOLDS AND ROPES

605.1 Handholds. Handholds shall be provided in accordance with Section 324.

Exception: Handholds shall not be provided for wave action and action rivers.

605.2 Rope and float line. A rope and float line shall be provided for all of the following situations:

1. Separation of activity areas.
2. Identification of a break in floor slope at water depths of less than 5 feet (1524 mm).
3. Identification of a water depth greater than $4^1/_2$ feet (1372 mm) in constant floor slope in Class D-2 pools.

Exception: Class D-1 pools or any other pool where the designer indicates that such a line is not required or that the line would constitute a hazard.

605.2.1 Location. The rope and float line shall be located 1 foot (305 mm) toward the shallow end in each location.

605.3 Caisson wall rope and float line. For Class D-1 pools, a rope and float line shall be installed to restrict bather access to the wave pool caisson wall. The location of the rope and float line shall be in accordance with the wave equipment manufacturer's instructions.

605.4 Fastening. Rope and float lines shall be securely fastened to wall anchors made of corrosion-resistant materials. Wall anchors shall be of the recessed type and shall not have projections that will constitute a hazard when the rope and float line is removed.

605.5 Size. Rope and float lines shall be not less than $^5/_8$ inch (15.9 mm) in diameter and shall be made of polypropylene material.

SECTION 606—DEPTHS

606.1 Class D-6 depth. The captured or standing water depth in Class D-6 pools shall be not greater than 12 inches (305 mm).

606.2 Spray pools. The water depths in spray pools shall be not greater than 6 inches (152 mm).

SECTION 607—BARRIERS

607.1 Barriers. Multiple pools and spas within a single complex shall be permitted without barriers where a barrier separates the single complex from the surrounding property in accordance with Section 305.

SECTION 608—NUMBER OF OCCUPANTS

608.1 Occupant load. The occupant load for the pools or spas in the facility shall be calculated in accordance with Table 608.1. The occupant load shall be the combined total of the number of users based on the pool or spa water surface area and the deck area surrounding the pool or spa. The deck area occupant load shall be based on the occupant load calculated where a deck is provided or based on an assumed 4-foot-wide (1219 mm) deck surrounding the entire perimeter of the pool or spa, whichever is greater.

TABLE 608.1—OCCUPANT LOAD				
	SHALLOW OR WADING AREAS	**DEEP AREA (NOT INCLUDING THE DIVING AREA)**	**DIVING AREA (PER EACH DIVING BOARD)**	**DECK AREA**
Vessel water surface area	8 sq. ft per user	10 sq. ft. per user	300 sq. ft. per user	—
Deck area	—	—	—	1 user per 15 sq. ft.
For SI: 1 square foot = 0.0929 m².				

608.2 Facility capacity. For multiple pools and spas in a single aquatic recreation facility, the total facility occupant capacity shall not be limited by the number of occupants calculated in accordance with Section 608.1.

SECTION 609—DRESSING AND SANITARY FACILITIES

609.1 General. Dressing and sanitary facilities shall be provided in accordance with the minimum requirements of the *International Building Code* and *International Plumbing Code* and Sections 609.2 through 609.9.

609.2 Number of fixtures. The minimum number of required water closets, urinals, lavatory, and drinking fountain fixtures shall be provided as required by the *International Building Code* and *International Plumbing Code* and the dressing facilities and number of cleansing and rinse showers shall be provided in accordance with Sections 609.2.1, 609.2.2, and 609.3.1.

609.2.1 Water area less than 7500 square feet. Facilities that have less than 7500 gross square feet (697 m²) of water area available for bather access shall have dressing facilities and not less than one cleansing shower for males and one cleansing shower for females.

Exception: This requirement shall not apply to Class C semipublic pools.

609.2.2 Water area 7500 square feet or more. Facilities that have 7500 gross square feet (697 m²) or more of water area available for bather access shall have dressing facilities and not less than one cleansing shower for males, and one cleansing shower for females for every 7500 square feet (697 m²) or portion thereof. Where the result of the fixture calculation is a portion of a whole number, the result shall be rounded up to the nearest whole number.

609.3 Showers. Showers shall be in accordance with Sections 609.3.1 through 609.3.3.

609.3.1 Rinse shower. In addition to the requirement for cleansing showers in Sections 609.2.1 and 609.2.2, not less than one rinse shower shall be provided on the deck of or at the entrance of each pool.

609.3.2 Water heater and mixing valve. Bather access to water heaters and thermostatically controlled mixing valves for showers shall be prohibited.

609.3.3 Temperature. At each cleansing showerhead, the heated shower water temperature shall be not less than 90°F (32°C) and not greater than 120°F (49°C). Water supplied to rinse showers shall not be required to be heated.

609.4 Soap dispensers. Soap dispensers shall be in accordance with Sections 609.4.1 and 609.4.2.

609.4.1 Liquid or powder. Soap dispensers shall be provided at each lavatory and cleansing shower. Soap dispensers shall dispense liquid or powdered soap. Reusable cake soap is prohibited. Soap dispensers and soap shall not be provided at rinse showers.

609.4.2 Metal or plastic. Soap dispensers shall be made of metal or plastic. Glass materials shall be prohibited.

609.5 Toilet tissue holder. A toilet paper holder shall be provided at each water closet.

609.6 Lavatory mirror. Where mirrors are provided, they shall be shatter resistant.

609.7 Sanitary napkin receptacles. Sanitary napkin receptacles shall be provided in each water closet compartment for females and in the cleansing area of the showers for female use only.

609.8 Sanitary napkin dispensers. A sanitary napkin dispenser shall be provided in each toilet facility for females.

609.9 Infant care. Baby-changing tables shall be provided in toilet facilities having two or more water closets.

SECTION 610—SPECIAL FEATURES

610.1 Locations. Entry and exit locations shall be in accordance with Table 610.1. The primary means of entry and exit shall consist of ramps, beach entries, pool stairs, or ladders.

TABLE 610.1—ENTRY AND EXIT LOCATIONS	
CLASS OF POOL	**ENTRY AND EXIT LOCATIONS**
D-1	Entry at beach end only; exit at beach end, sides or end wall
D-2	Entry and exit determined by the pool designer
D-3	Entry prohibited from deck areas; exit by ladders, steps or ramps as determined by pool designer

TABLE 610.1—ENTRY AND EXIT LOCATIONS—continued

CLASS OF POOL	ENTRY AND EXIT LOCATIONS
D-4	Entry and exit determined by the pool designer
D-5	Entry and exit determined by the pool designer
D-6	Entry and exit determined by the pool designer

610.2 Secondary entry and exit means. Where secondary means of entry and exit are provided, they shall consist of one of the following:

1. Steps.
2. Stairs.
3. Ladders with grab rails.
4. Recessed treads.
5. Ramps.
6. Beach entries.
7. Swimouts.
8. Designs that provide the minimum utility as specified in this code.

610.3 Provisions for diving. Where diving facilities are part of the attraction or pool complex, entries, exits, pool stairs, ladders, underwater benches, special features, and other accessories shall be located outside of the minimum diving water envelope in accordance with Figure 323.2.

610.4 Beach entry, zero-depth entry, and sloping entries. The shallow end for beach entries and sloping entries shall be in accordance with Sections 610.4.1 through 610.4.4 or the regulations of the local jurisdiction.

610.4.1 Maximum entry slope. The slope of sloping entries used as a pool entry shall not exceed 1 unit vertical in 12 units horizontal (8.3-percent slope).

610.4.2 Benches. Where benches are used in conjunction with sloping entries, the vertical riser height shall not exceed 12 inches (305 mm).

610.4.3 Steps. Where steps are used in conjunction with sloping entries, all of the requirements of Section 610.5 shall apply.

610.4.4 Slip-resistant surfaces. Beach and sloping entry walking surfaces at water depths up to 36 inches (914 mm) shall be slip resistant.

610.5 Pool steps. The design and construction of steps for stairs into the shallow end and recessed pool stairs shall be in accordance with Sections 610.5.1 through 610.5.6.

610.5.1 Uniform height of 9 inches. Except for the bottom riser, risers at the centerline shall have a maximum uniform height of 9 inches (229 mm). The bottom riser height shall be permitted to vary from the other risers.

610.5.2 Distance from coping or deck. The vertical distance from the pool coping, deck, or step surface to the uppermost tread shall be not greater than 9 inches (229 mm).

610.5.3 Color to mark leading edge. The leading edge of steps shall be distinguished by a color contrasting with the color of the steps and the pool floor.

610.5.4 Stairs in water depths over 48 inches. Stairs that are located in water depths greater than 48 inches (1219 mm) shall have the lowest tread located below the deck at a distance of not less than 48 inches (1219 mm) below the deck.

610.5.5 Tread horizontal depth. Treads shall have an unobstructed horizontal depth of not less than 11 inches (279 mm).

610.5.6 Tread surface area. Treads shall have an unobstructed surface area of not less than 240 square inches (.017 m²).

610.6 Swimouts. Swimouts shall be located completely outside of the water current or wave action of the pool or spa and can be located in shallow or deep areas of water.

610.6.1 Surface area. An unobstructed surface equal to or greater than that required for the top tread of the pool stairs shall be provided in accordance with Sections 610.5.5 and 610.5.6.

610.6.2 Step required. Where a swimout is used as an entry and exit access point, it shall be provided with a step that meets the pool stair requirements (see Section 610.5).

610.6.3 Maximum depth. The horizontal surface of a swimout shall be not greater than 20 inches (508 mm) below the waterline.

610.6.4 Color marking. The leading edge of a swimout shall be visually set apart by a stripe having a width of not less than $^3/_4$ inch (19 mm) and not greater than 2 inches (51 mm). The stripe shall be of a contrasting color to the adjacent surfaces.

610.7 Underwater benches. Underwater benches shall comply with this section.

610.7.1 Location. Underwater benches shall only be located in areas where the pool water depth does not exceed 5 feet (1.5 m).

610.7.2 Surface dimensions. Underwater benches shall have an unobstructed surface dimension of not less than 16 inches (406 mm) and not greater than 22 inches (559 mm) in depth measured front to back and not less than 26 inches (660 mm) in width.

610.7.3 Not an entry or exit. Underwater benches shall not be used as an entry or exit for a pool.

610.7.4 Depth. The horizontal surface of benches shall be not greater than 20 inches (508 mm) below the waterline.

610.7.5 Color marking. The leading edge of benches shall be visually set apart by a stripe having a width not less than $^3/_4$ inch (19 mm) and not greater than 2 inches (51 mm). The stripe shall be of a contrasting color to the adjacent surfaces.

610.7.6 Slip resistant. The top surface of benches shall be slip resistant.

610.8 Objects permitted. The design, construction, and operation of decorative objects and structures intended for climbing, walking, and hanging on by a bather are not covered by this code.

610.8.1 Floating devices. Floating devices not intended to be mobile shall be anchored in a manner to restrict movement to the range established by the designer. The anchoring of such floating devices shall be configured to minimize the possibility of entrapment of bathers, bodies, hair, limbs, and appendages should they come in contact with any element of the floating device or its anchors.

SECTION 611—SIGNAGE

611.1 Posting of signs. Signs stating rules, instructions, and warnings shall be posted. Signs for suction entrapment warning in accordance with Section 311 shall be posted. Signs shall be placed so that they squarely face approaching traffic. The center of the message panel shall be located not less than 66 inches (1676 mm) above the walking surface.

611.2 Prohibited mounting. Signs shall not be mounted on fences and gates alongside of guest walkways and staircases.

611.3 Message delivery. Messages delivered on signs shall comply with all of the following:

1. Messages shall be pertinent to the activity being performed or to be performed.
2. Messages shall be specific by providing details about the activity.
3. Messages shall be short and concise.
4. Messages shall be direct without humor or embellishments.

611.4 Text font and size. The message text shall be in a clear, bold font such as Arial. The character height shall be proportional to 1 inch (25 mm) for 10 feet (3048 mm) of intended viewing distance but not less than 1 inch (25 mm).

611.5 Distinct sign classes. Facility signs shall be categorized into four sign classes in accordance with Sections 611.5.1 through 611.5.4.

611.5.1 General information. General information signs shall be posted facility-wide and shall not be attraction specific.

611.5.2 Directional signs. Directional signs shall identify the location of services and attractions in the park and shall include directional arrows. Directional signs shall be posted at various crossroads in the facility.

611.5.3 Rule signs. Rule signs shall inform guests of the qualifications that they must meet to allow them to participate on a specific ride or attraction. Rules shall include, but are not limited to, limits for weight and height, proper attire and ride (and ride vehicle) stipulations. Rule signs shall be located at a point where the guests make the initial commitment to participate on the ride.

611.5.4 Instructional signs. Instructional signs shall inform guests of specific instructions for the use of the ride. Instructions shall include, but are not limited to, riding posture, prohibited activity, and user exit requirements at the ride termination. Instructional signs shall be located along the queue approaching the ride dispatch area.

611.6 Materials. Sign panels shall be durable for the weather conditions and shall be resistant to damage from guests. The message surface shall be clean and smooth and shall readily accept paint or precut lettering adhesives.

611.7 Shape and size consistency. The panel shape and size for each class of signs shall be the same. Where the total message to be indicated is larger than what can be placed on one sign, multiple signs of the same size shall be used to display the message.

611.8 Pictograms. Pictograms shall always be accompanied by text indicating the same message. Pictograms shall be designed to illustrate one clear and specific meaning to all individuals.

611.9 Theming or artwork. Theming or artwork applied to signs shall not invade the message panel. Signs shall have a distinct border.

611.10 Shallow water. Safety signs shall be in accordance with Section 412.

611.11 Cold water. Where a pool could have a water temperature below 70°F (21°C), a cold water warning sign shall be posted at the point of entry to the pool or at the attraction using such water.

SECTION 612—INTERACTIVE WATER PLAY FEATURES

612.1 General. Interactive water play features shall comply with Sections 612.1 through 612.7.

612.2 Safety hazards. Parts of the interactive water play feature that can be accessed by the users of the feature shall be designed and constructed to not present safety hazards to the users.

612.3 Decking. A deck of not less than 4 feet (1296 mm) in width shall be provided around the perimeter of the interactive water play feature. The deck shall be sloped away from the interactive water play feature.

612.4 Splash pad zone. The splash pad zone shall comply with Sections 612.4.1 through 612.4.4.

612.4.1 Surface. Splash pad zone surfaces shall have a slip-resistant and cleanable surface. The manufacturer of manufactured zone surfaces shall certify that such surfacing is suitable for aquatic and chlorinated environments. Direct suction outlets from interactive water play features shall be prohibited.

612.4.2 Slope and water collection. Splash pad zone surfaces shall slope to one or more drain points so that only water from the splash pad zone flows back to a gravity-fed collection tank. The slope shall prevent the accumulation or pooling of water and shall not exceed $^1/_2$ inch per foot (41.6 mm/m).

Drain openings in the splash pad zone surfaces that can be accessed by users shall not allow a $^1/_2$-inch (12.7 mm) diameter dowel rod to be inserted into the opening. Drain covers in splash pad zone surfaces shall be flat and flush with the zone surface and shall require tools for removal. The manufacturer of such drain covers shall certify that the covers comply with the physical testing and finger-and-limb entrapment requirements in Sections 3 and 6, respectively, of APSP 16.

612.4.3 Nozzles within the interactive water play feature splash pad zone. Nozzles that spray water from the interactive water play feature splash pad zone shall be flush with the zone surface. Openings in such nozzles shall not allow a $^1/_2$ inch (12.7 mm) diameter dowel rod to be inserted into the opening. The water velocity from the orifice of any water nozzle shall not exceed 20 feet (6.1 m) per second.

612.4.4 Other nozzles. Nozzles, other than those on walking surfaces within the interactive water play feature splash pad zone, shall be designed to be clearly visible.

612.5 Water sanitation. The water sanitation shall consist of the equipment covered in Sections 612.5.1 through 612.5.5.

612.5.1 Water collection and treatment tank. Interactive water play features shall drain to a collection and treatment tank. The inside of the tank shall be provided access for cleaning and inspection. The access hatch or lid shall be locked or require a tool to open.

612.5.2 Filtration pump. The filtration pump shall be sized to turn over the surge basin capacity in 30 minutes or less. The intake for the pump shall be located to draw water from the lowest elevation in the treatment tank.

612.5.3 Spray nozzle and water feature water disinfection. Spray nozzles and water features shall be supplied by water from the water collection and treatment tank that is equipped with filtration and sanitizing equipment required by Chapter 3 and this section. Where separate water feature pumps are installed, controls shall prevent those pumps from operation when the filtration pump is not operating.

612.5.4 Disinfection system. In addition to any filtration and sanitizing equipment requirements of Chapter 3 and this chapter, all water supplied to spray nozzles or other water that can be accessed by a user shall pass through a secondary disinfection system before discharge to the user. The secondary disinfection system shall be *listed* and *labeled* to NSF 50 as having a single-pass, three-log reduction of the cryptosporidium surrogate.

612.5.5 Make-up water system. The water collection and treatment tank shall be provided with a make-up water system that is supplied with potable water.

612.6 Operating instructions. In addition to the documentation and instructions requirements of Chapters 1 and 3, the operating instructions for an interactive water play feature shall require that the circulation system be operated continuously for not less than four turnovers prior to operation of the pumps for the spray nozzles and other water features systems.

612.7 Lighting. Where a interactive water play feature will be in operation at night or during periods of inadequate natural lighting, artificial lighting shall be provided in accordance with the same requirements for pool deck area lighting in Section 322.2.1.

ONGROUND STORABLE RESIDENTIAL SWIMMING POOLS

User notes:

About this chapter: *Chapter 7 concerns residential portable pools also known as onground storable residential swimming pools. These pools are manufactured for assembly on the site. The chapter's regulations include those for floor slopes, entry barrier methods, decks, stairs, safety signage and circulation systems.*

SECTION 701—GENERAL

701.1 Scope. This chapter describes certain criteria for the design, manufacturing, and testing of *onground storable pools* intended for *residential* use. This includes portable pools with flexible or nonrigid side walls that achieve their structural integrity by means of uniform shape, support frame or a combination thereof, and that can be disassembled for storage or relocation. This chapter includes what has been commonly referred to in past standards or codes as onground or above-ground pools.

    701.1.1 Permanent inground residential swimming pool. This chapter does not apply to permanent inground *residential* pools, as defined in Chapter 8.

701.2 General. In addition to the requirements of this chapter, onground storable *residential* swimming pools shall comply with the requirements of Chapter 3.

701.3 Floor slopes. Floor slopes shall be uniform and in accordance with Sections 701.3.1 through 701.3.4.

    701.3.1 Shallow end. The slope of the floor from the shallow end wall towards the deep area shall not exceed 1 unit vertical in 7 units horizontal (14-percent slope) to the point of the first slope change.

    701.3.2 Transition. The slope of the floor from the point of the first slope change towards the deepest point shall not exceed 1 unit vertical in 3 units horizontal (33-percent slope).

    701.3.3 Adjacent. The slope adjacent to the shallow area shall not exceed 1 unit vertical in 3 units horizontal (33-percent slope) and the slope adjacent to the side walls shall not exceed 1 unit vertical in 1 unit horizontal (100-percent slope).

    701.3.4 Change point. The point of the first slope change shall be defined as the point at which the shallow area slope exceeds 1 unit vertical in 7 units horizontal (14-percent slope) and is not less than 6 feet (1889 mm) from the shallow end wall of the pool.

701.4 Identification. For onground storable *residential* pools with a vinyl liner, the manufacturer's name and the liner identification number shall be affixed to the liner. For onground storable *residential* pools without a liner, the manufacturer's name and identification number shall be affixed to the exterior of the pool structure.

701.5 Installation. *Onground storable pools* shall be installed in accordance with the manufacturer's instructions.

SECTION 702—LADDERS AND STAIRS

702.1 Ladders and stairs. Pools shall have a means of entry and exit consisting of not less than one ladder or a ladder and staircase combination.

702.2 Type A and Type B ladders. Type A, double access, and Type B, limited access, A-frame ladders shall comply with Sections 702.2.1 through 702.2.7. See Figure 702.2.

FIGURE 702.2—TYPICAL A-FRAME LADDER, TYPES A AND B

    702.2.1 Barrier required. Ladders in the pool shall have a physical barrier to prevent children from swimming through the riser openings or behind the ladder.

    Exception: Barriers for ladders shall not be required where the ladder manufacturer provides a certification statement that the ladder complies with the ladder entrapment test requirements of APSP 4.

702.2.2 Platform. Where an A-frame ladder has a platform between the handrails, the platform shall have a width of not less than 12 inches (305 mm) and a length of not less than 12 inches (305 mm). The platform shall be at or above the highest ladder tread. The walking surface of the platform shall be slip resistant.

702.2.3 Handrails or handholds. A-frame ladders shall have two handrails or handholds that serve all treads. The height of the handrails and handholds shall be not less than 20 inches (508 mm) above the platform or uppermost tread, whichever is higher.

702.2.4 Diameter. The outside diameter of handrails and handholds shall be not less than 1 inch (25 mm) and not greater than 1.9 inches (48 mm).

702.2.5 Clear distance. The clear distance between ladder handrails shall be not less than a space of 12 inches (305 mm).

702.2.6 Treads. Ladder treads shall have a horizontal uniform depth of not less than 2 inches (51 mm).

702.2.7 Riser height. Risers, other than the bottom riser, shall be of uniform height that is not less than 7 inches (178 mm) and not greater than 12 inches (305 mm). The bottom riser height shall be not less than 7 inches (178 mm) and not greater than 12 inches (305 mm). The vertical distance from the platform or top of the pool structure to the uppermost tread shall be the same as the uniform riser heights.

702.3 Type C staircase ladders (ground to deck). Type C staircase ladders shall comply with Sections 702.3.1 through 702.3.6. See Figure 702.3.

FIGURE 702.3—TYPICAL STAIRCASE LADDER, TYPE C

702.3.1 Handrails or handholds. Staircase ladders shall have not less than two handrails or handholds that serve all treads. The height of the handrails and handholds shall be not less than 20 inches (508 mm) above the platform or uppermost tread, whichever is higher.

702.3.2 Diameter. The outside diameter of handrails and handholds shall be not less than 1 inch (25 mm) and not greater than 1.9 inches (48 mm).

702.3.3 Treads. Ladder treads shall have a horizontal uniform depth of not less than 4 inches (102 mm).

702.3.4 Riser height. Risers, other than the bottom riser, shall be of uniform height that is not less than 7 inches (178 mm) and not greater than 12 inches (305 mm). The bottom riser height shall be not less than 7 inches (178 mm) and not greater than 12 inches (305 mm). The vertical distance from the platform or top of the pool structure to the uppermost tread shall be the same as the uniform riser heights.

702.3.5 Top step. The top step of a staircase ladder shall be flush with the deck or 7 inches (178 mm) to 12 inches (305 mm) below the deck level.

702.3.6 Width. Steps shall have a minimum unobstructed width of 19 inches (483 mm) between the side rails.

702.4 Type D in-pool ladders. Type D in-pool ladders shall be in accordance with Sections 702.4.1 through 702.4.7. See Figure 702.4.

FIGURE 702.4—TYPICAL IN-POOL LADDER, TYPE D

702.4.1 Clearance. There shall be a clearance of not less than 3 inches (76 mm) and not greater than 6 inches (152 mm) between the pool wall and the ladder.

702.4.2 Handrails or handholds. Ladders shall be equipped with two handrails or handholds that extend above the platform or deck not less than 20 inches (508 mm).

702.4.3 Clear distance. The clear distance between ladder handrails shall be not less than 12 inches (305 mm).

702.4.4 Diameter. The outside diameter of handrails and handholds shall be not less than 1 inch (25 mm) and not greater than 1.9 inches (48 mm).

702.4.5 Riser height. Risers, other than the bottom riser, shall be of uniform height that is not less than 7 inches (178 mm) and not greater than 12 inches (305 mm). The bottom riser height shall be not less than 7 inches (178 mm) and not greater than 12 inches (305 mm).

702.4.6 Top tread. The vertical distance from the pool coping, deck, or step surface to the uppermost tread shall be not less than 7 inches (178 mm) and not greater than 12 inches (305 mm) and uniform with other riser heights.

702.4.7 Tread depth. Ladder treads shall have a horizontal uniform depth of not less than 2 inches (51 mm).

702.5 Type E protruding in-pool stairs. Type E protruding in-pool stairs shall be in accordance with Sections 702.5.1 through 702.5.7. See Figure 702.5.

FIGURE 702.5—TYPICAL IN-POOL STAIRCASE TYPES, E AND F

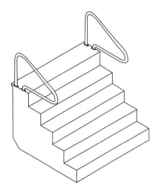

702.5.1 Barrier required. In-pool stairs shall have a physical barrier to prevent children from swimming through the riser openings or behind the in-pool stairs.

702.5.2 Handrails or handholds. In-pool stairs shall be equipped with not less than one handrail or handhold that serves all treads with a height of not less than 20 inches (508 mm) above the platform or uppermost tread, whichever is higher.

702.5.3 Removable handrails. Where handrails are removable, they shall be installed such that they cannot be removed without the use of tools.

702.5.4 Leading edge distance. The leading edge of handrails shall be 18 inches (457 mm) ± 3 inches (± 76 mm), horizontally from the vertical plane of the bottom riser.

702.5.5 Diameter. The outside diameter of handrails and handholds shall be not less than 1 inch (25 mm) and not greater than 1.9 inches (48 mm).

702.5.6 Tread width and depth. Treads shall have an unobstructed horizontal depth of not less than 10 inches (254 mm) and an unobstructed surface area of not less than 240 square inches (0.15 m^2).

702.5.7 Uniform riser height. Risers, other than the bottom riser, shall be of uniform height that is not less than 7 inches (178 mm) and not greater than 12 inches (305 mm). The bottom riser height shall be not less than 7 inches (178 mm) and not greater than 12 inches (305 mm). The vertical distance from the pool coping, deck or step surface to the uppermost tread of the stairs shall be the same as the uniform riser heights.

702.6 Type F recessed in-pool stairs. Type F recessed in-pool stairs shall be in accordance with Sections 702.6.1 through 702.6.7. See Figure 702.5.

702.6.1 Barrier required. In-pool stairs shall have a physical barrier to prevent children from swimming through the riser openings or behind the in-pool stairs.

702.6.2 Handrails or handholds. In-pool stairs shall be equipped with not less than one handrail or handhold that serves all treads with a height of not less than 20 inches (508 mm) above the platform or uppermost tread, whichever is higher.

702.6.3 Removable handrails. Where handrails are removable, they shall be installed such that they cannot be removed without the use of tools.

702.6.4 Leading edge distance. The leading edge of handrails shall be 18 inches (457 mm) ± 3 inches (± 76 mm), horizontally from the vertical plane of the bottom riser.

702.6.5 Diameter. The outside diameter of handrails and handholds shall be not less than 1 inch (25 mm) and not greater than 1.9 inches (48 mm).

702.6.6 Tread width and depth. Treads shall have an unobstructed horizontal depth of not less than 10 inches (254 mm) at all points and an unobstructed surface area of not less than 240 square inches (0.15 m^2).

702.6.7 Uniform riser height. Risers, other than the bottom riser, shall be of uniform height that is not less than 7 inches (178 mm) and not greater than 12 inches (305 mm). The bottom riser height shall be not less than 7 inches (178 mm) and not greater than 12 inches (305 mm). The vertical distance from the pool coping, deck or step surface to the uppermost tread of the stairs shall be the same as the uniform riser heights.

SECTION 703—DECKS

703.1 General. Decks provided by the pool manufacturer shall be installed in accordance with the manufacturer's instructions. Decks fabricated on-site shall be in accordance with the *International Residential Code*.

703.2 Cantilevered. The top surface of a cantilevered deck shall be not greater than 1 inch (25 mm) higher than the top of the pool wall. See Figure 703.4. The top surface of a noncantilevered deck shall be not higher than the top of the pool wall.

703.3 No gaps. Decks that are installed flush with the top rail of the pool shall have all gap openings between the deck and top rails closed-off or capped.

703.4 Extension over pool. Where a deck extends inside the top rail of the pool, it shall extend not more than 3 inches (76 mm) beyond the inside of the top rail of the pool in accordance with Figure 703.4 and shall have a smooth finish.

FIGURE 703.4—TYPICAL CANTILEVERED DECK SUPPORT

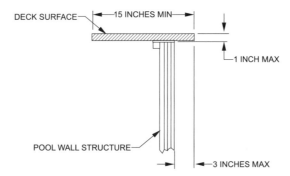

703.5 Slip resistant. The deck walking surface shall be slip resistant.

703.6 Walk-around decks. Walk-around decks shall have a level walking surface of not less than 15 inches (381 mm) in width, as measured from the inside edge of the pool top rail to the outside of the pool walk-around. See Figure 703.6.

FIGURE 703.6—WALK-AROUND DECK WIDTH

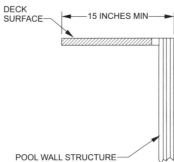

SECTION 704—CIRCULATION SYSTEM

704.1 General. A circulation system consisting of pumps, hoses, tubing, piping, return inlets, suction outlets, filters and other related equipment that provides for the circulation of water throughout the pool shall be located so that such items cannot be used by young children as a means of access to the pool.

704.2 Installation and support. Circulation equipment shall be installed, mounted and supported in accordance with the manufacturer's instructions.

704.3 Draining the system. In climates subject to freezing, circulation system equipment shall be designed and fabricated to drain the pool water from the equipment and exposed piping, by removal of drain plugs and manipulating valves or by other methods in accordance with the manufacturer's instructions.

704.4 Turnover. A pump including a motor shall be provided for circulation of the pool water. The equipment shall be sized to provide a turnover of the pool water not less than once every 12 hours. The system shall be designed to provide the required turnover rate based on the manufacturer's specified maximum flow rate of the filter, with a clean media condition of the filter. The system flow shall not exceed the filter manufacturer's maximum filter flow rate.

704.5 Piping and fittings. The process piping of the circulation system, including but not limited to hoses, tubing, piping, and fittings, shall be made of nontoxic material and shall be capable of withstanding an internal pressure of not less than $1^1/_2$ times the rated pressure of the pump. Piping on the suction side of the pump shall not collapse when flow into such piping is blocked.

704.6 Filters. Pressure-type filters shall have an automatic internal means or a manual external means to relieve accumulated air pressure inside the filter tank. Filter tanks composed of upper and lower tank lids that are held in place by a perimeter clamp shall have a perimeter clamp that provides for a slow and safe release of air pressure before the clamp disengages the lids.

704.6.1 Automatic internal air relief. Filter tanks incorporating an automatic internal air relief as the principal means of air release shall be designed with a means to provide for a slow and safe release of pressure.

704.6.2 Separation tank. A separation tank used in conjunction with a filter tank shall have a manual air release or the tank shall be designed to provide for a slow and safe release of pressure when the tank is opened.

704.7 Pumps. Pool pumps shall be tested and certified by a nationally recognized testing laboratory in accordance with UL 1081.

704.7.1 Cleanable strainer. Where a pressure-type filter is installed, a cleanable strainer or screen that captures materials such as solids, debris, hair and lint shall be provided upstream of the circulation pump.

704.7.2 Access to pumps and motors. Pumps and motors shall be provided access for inspection and service in accordance with the pump and motor manufacturer's instructions.

704.7.3 Pump shutoff valves. An available means of shutting off of the suction and discharge piping for the pump shall be provided for maintenance and removal of the pump and be located with access.

704.8 Outlets and return inlets. Outlets or suction outlets and return inlets shall be provided and arranged to produce uniform circulation of water so that sanitizer residual is maintained throughout the pool. Where installed, submerged suction outlets shall conform to APSP 16.

704.9 Surface skimmer systems. The surface skimming system provided shall be designed and constructed to skim the pool surface where the water level is maintained between the minimum and maximum fill level of the pool.

704.9.1 Coverage where used as a sole outlet. Where surface skimmers are used as the only pool water outlet system, not less than one skimmer shall be provided for each 800 square feet (74.3 m^2), or fraction thereof, of the water surface area.

704.9.2 Coverage where used in combination with other outlets. Where surface skimmers are not the only outlet for pool water, they shall be considered to cover only that fraction of the 800 square feet (74.3 m^2).

704.9.3 Location and venting. Skimmers shall be equipped with a vent that serves as a vacuum break.

SECTION 705—SAFETY SIGNS

705.1. Signs to be installed prior to final inspection. Safety signage such as "NO DIVING" signs and other safe use instruction signs that are provided by the pool and ladder manufacturer shall be posted in accordance with the manufacturer's instructions prior to final inspection.

705.2 Safety signs for ladders. Safety signage for ladders shall be in accordance with Sections 705.2.1 through 705.2.3.2.

705.2.1 A-frame ladders. Safety signage for A-frame ladders shall be in accordance with Sections 705.2.1.1 through 705.2.1.4.1. The words on the signage shall be readable by persons standing in the pool and standing outside of the pool as applicable for the required location of each sign.

705.2.1.1 No diving warning. A-frame ladders shall have the following words posted on the in-pool side of the ladder and on the pool entry side of the ladder: "NO DIVING." The location of the words shall be above the elevation of the design water level of the pool.

705.2.1.2 Entrapment warning. A-frame ladders shall have the following words posted on the pool side of the ladder: "TO PREVENT ENTRAPMENT OR DROWNING DO NOT SWIM THROUGH, BEHIND, OR AROUND LADDER."

705.2.1.3 Type A, A-frame ladders. Type A double access A-frame ladders shall have the following words posted on the ladder: "REMOVE AND SECURE LADDER WHEN POOL IS NOT OCCUPIED."

705.2.1.4 Type B, A-frame ladders. Type B limited access A-frame ladders shall have the following words posted on the ladder: "SECURE LADDER WHEN POOL IS NOT OCCUPIED."

705.2.1.4.1 Swing up or slide up secured ladders. Type B limited access A-frame ladders that utilize swing-up or slide-up sections for limiting access to the pool shall have the following words posted on the ladder as applicable for the type of securing method:

1. "WHEN POOL IS NOT OCCUPIED, SWING UP AND SECURE."
2. "WHEN POOL IS NOT OCCUPIED, LIFT OFF."
3. "WHEN POOL IS NOT OCCUPIED, SLIDE UP AND SECURE."

705.2.2 Type C staircase ladders. Type C staircase ladders that swing up to limit access to the pool or that are removed to limit access to the pool shall have the following words posted on the ladder: "WHEN NOT IN USE SWING UP AND SECURE OR REMOVE."

705.2.3 Type D in-pool ladder. Safety signage for Type D in-pool ladders shall be in accordance with Sections 705.2.3.1 and 705.2.3.2. The words on the signage shall be readable by persons standing in the pool or standing outside the pool as applicable for the required location of each sign.

705.2.3.1 No diving warning. Type D in-pool ladders shall have the following words posted on the in-pool side of the ladder and on the pool entry side of the ladder: "NO DIVING." The location of the words shall be above the elevation of the design water level of the pool.

705.2.3.2 Entrapment warning. Type D in-pool ladders shall have the following words posted on the ladder: "WARNING: TO PREVENT ENTRAPMENT OR DROWNING, DO NOT SWIM THROUGH, BEHIND, OR AROUND LADDER."

PERMANENT INGROUND RESIDENTIAL SWIMMING POOLS

User notes:

About this chapter: *Permanent inground residential swimming pools are regulated by Chapter 8. Where diving boards are present, this chapter provides information regarding the minimum diving water dimensions. Requirements for means of entry and exit, decks and circulation systems are provided. Special features of these pools such as beach entries, swimouts, diving rocks and architectural features are also regulated by this chapter.*

SECTION 801—GENERAL

801.1 Scope. The provisions of this chapter shall govern permanent inground *residential* swimming pools. Permanent inground *residential* swimming pools shall include pools that are partially or entirely above grade. This chapter does not cover pools that are specifically manufactured for above-ground use and that are capable of being disassembled and stored. This chapter covers new construction, modification and repair of inground *residential* swimming pools.

801.2 General. Permanent inground *residential* pools shall comply with the requirements of Chapter 3.

SECTION 802—DESIGN

802.1 Materials of components and accessories. The materials of components and accessories used for permanent inground *residential* swimming pools shall be suitable for the environment in which they are installed. The materials shall be capable of fulfilling the design, installation and the intended use requirements in the *International Residential Code.*

802.2 Structural design. The structural design and materials shall be in accordance with the *International Residential Code.*

SECTION 803—CONSTRUCTION TOLERANCES

803.1 Construction tolerances. The construction tolerance for dimensions for the overall length, width and depth of the pool shall be ± 3 inches (76 mm). The construction tolerance for all other dimensions, except the location of the *design waterline*, shall be ± 2 inches (51 mm), unless otherwise specified by the design engineer. The construction tolerance for the location of the *design waterline* shall be in accordance with Table 803.1.

TABLE 803.1—DESIGN WATERLINE CONSTRUCTION TOLERANCE

SURFACE	TOLERANCE
Waterline on tiled surface	± $^1/_4$ inch
Waterline on surfaces other than a tiled surface	± $^1/_2$ inch

SECTION 804—DIVING WATER ENVELOPES

804.1 General. The minimum diving water envelopes shall be in accordance with Table 804.1 and Figure 804.1. Negative construction tolerances shall not be applied to the dimensions of the minimum diving water envelopes given in Table 804.1.

TABLE 804.1—MINIMUM DIVING WATER ENVELOPE FOR SWIMMING POOLS DESIGNATED TYPES I-V[b]

POOL TYPE	MINIMUM DEPTHS AT POINT FEET-INCHES				MINIMUM WIDTHS AT POINT FEET-INCHES				MINIMUM LENGTHS BETWEEN POINTS FEET-INCHES					
	A	B	C	D	A	B	C	D	WA	AB	BC	CD	DE	WE
I	6-0	7-6	5-0	2-9	10-0	12-0	10-0	8-0	1-6	7-0	7-6	Note a	6-0	28-9
II	6-0	7-6	5-0	2-9	12-0	15-0	12-0	8-0	1-6	7-0	7-6	Note a	6-0	28-9
III	6-0	8-0	5-0	2-9	12-0	15-0	12-0	8-0	2-0	7-6	9-0	Note a	6-0	31-3
IV	7-8	8-0	5-0	2-9	15-0	18-0	15-0	9-0	2-6	8-0	10-6	Note a	6-0	31-3

POOL TYPE	MINIMUM DEPTHS AT POINT FEET-INCHES				MINIMUM WIDTHS AT POINT FEET-INCHES				MINIMUM LENGTHS BETWEEN POINTS FEET-INCHES					
	A	B	C	D	A	B	C	D	WA	AB	BC	CD	DE	WE
V	8-6	9-0	5-0	2-9	15-0	18-0	15-0	9-0	3-0	9-0	12-0	Note a	6-0	36-9

TABLE 804.1—MINIMUM DIVING WATER ENVELOPE FOR SWIMMING POOLS DESIGNATED TYPES I-V[b]—continued

For SI: 1 inch = 25.4 mm, 1 foot = 304.8 mm.
a. The minimum length between points C and D varies based on water depth at point D and the floor slope between points C and D.
b. See Figure 804.1 for location of points.

FIGURE 804.1—MINIMUM DIVING WATER ENVELOPE

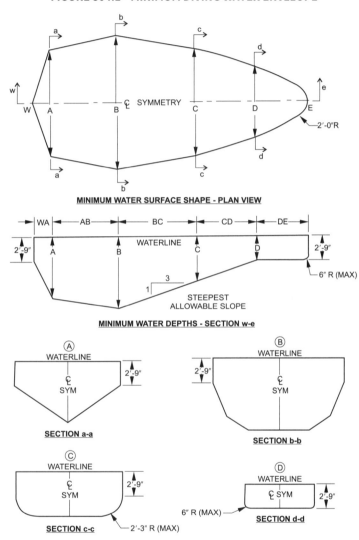

SECTION 805—WALLS

805.1 General. Walls in the shallow area and deep area of the pool shall have a wall-to-floor transition point that is not less than 33 inches (838 mm) below the *design waterline*. Above the transition point, the walls shall be within 11 degrees (0.19 rad) of vertical.

SECTION 806—OFFSET LEDGES

806.1 Maximum width. Offset ledges shall be not greater than 8 inches (203 mm) in width.

806.2 Reduced width required. Where an offset ledge is located less than 42 inches (1067 mm) below the *design waterline*, the width of such ledge shall be proportionately less than 8 inches (203 mm) in width so as to fall within 11 degrees of vertical as measured from the top of the *design waterline*.

SECTION 807—POOL FLOORS

807.1 Floor slopes. Floor slopes shall be in accordance with Sections 807.1.1 through 807.1.3.

807.1.1 Shallow end. The slope of the floor from the beginning of the shallow end to the deep area floor slope transition point, indicated in Figure 804.1 as Point E to Point D, shall not exceed 1 unit vertical in 7 units horizontal.

807.1.2 Shallow to deep transition. The shallow to deep area floor slope transition point, indicated in Figure 804.1 as Point D, shall occur at a depth not less than 33 inches (838 mm) below the *design waterline* and at a point not less than 6 feet (1829 mm) from the beginning of the shallow end, indicated in Figure 804.1 as Point E, except as specified in Section 809.7.

807.1.3 Deep end. The slope of the floor in the deep end, indicated in Figure 804.1 as Point B to Point D, shall not exceed a slope of 1 unit vertical in 3 units horizontal (33-percent slope).

807.2 Shallow end water depths. The design water depth as measured at the shallowest point in the shallow area shall be not less than 33 inches (838 mm) and not greater than 4 feet (1219 mm). Shallow areas designed in accordance with Sections 809.6, 809.7 and 809.8 shall be exempt from the minimum depth requirement.

SECTION 808—DIVING EQUIPMENT

808.1 Manufactured and fabricated diving equipment. Manufactured and fabricated diving equipment shall be in accordance with this section. Manufactured and fabricated diving equipment and appurtenances shall not be installed on a Type O pool.

808.2 Manufactured diving equipment. Manufactured diving equipment shall be designed for swimming pool use.

808.3 Installation. Where manufactured diving equipment is installed, the installation shall be located in the deep area of the pool so as to provide the minimum dimensions as shown in Table 804.1 and shall be installed in accordance with the manufacturer's instructions.

808.4 Labeling. Manufactured diving equipment shall have a permanently affixed label indicating the manufacturer's name and address, the date of manufacture, the minimum diving envelope and the maximum weight limitation.

808.5 Slip resistant. Diving equipment shall have slip-resistant walking surfaces.

808.6 Point A. For the application of Table 804.1, Point A shall be the point from which all dimensions of width, length and depth are established for the minimum diving water envelope. If the tip of the diving board or diving platform is located at a distance of WA or greater from the deep end wall and the water depth at that location is equal to or greater than the water depth requirement at Point A, then the point on the water surface directly below the center of the tip of the diving board or diving platform shall be identified as Point A.

808.7 Location of pool features in a diving pool. Where a pool is designed for use with diving equipment, the location of steps, pool stairs, ladders, underwater benches, special features and other accessory items shall be outside of the minimum diving water envelope as indicated in Figure 323.2.

808.8 Stationary diving platforms and diving rocks. Stationary diving platforms and diving rocks built on-site shall be permitted to be flush with the wall and shall be located in the diving area of the pool. Point A shall be in front of the wall at the platform or diving rock centerline.

808.9 Location. The forward tip of manufactured or fabricated diving equipment shall be located directly above Point A as defined by Section 808.6.

808.10 Elevation. The maximum elevation of a diving board above the *design waterline* shall be in accordance with the manufacturer's instructions.

808.11 Minimum water envelope. Manufactured diving equipment installation and use instructions shall be provided by the diving equipment manufacturer and shall specify the minimum water dimensions required for each diving board and diving stand combination. The board manufacturer shall indicate the water envelope type by dimensionally relating their products to Point A on the water envelopes as shown in Figure 804.1 and Table 804.1. The board manufacturer shall specify which boards fit on the design pool geometry types as indicated in Table 804.1.

808.12 Platform height above waterline. The height of a stationary diving platform or a diving rock above the *design waterline* shall not exceed the dimensions in Table 808.12.

TABLE 808.12—DIVING PLATFORM OR APPURTENANCE HEIGHT ABOVE DESIGN WATERLINE	
POOL TYPE	HEIGHT INCHES
I	42
II	42
III	50
IV	60
V	69
For SI: 1 inch = 25.4 mm.	

808.13 Headroom above the board. The diving equipment manufacturer shall specify the minimum headroom required above the board tip.

SECTION 809—SPECIAL FEATURES

809.1 Slides. Slides shall be installed in accordance with the manufacturer's instructions.

809.2 Entry and exit. Pools shall have a means of entry and exit in all shallow areas where the design water depth of the shallow area at the shallowest point exceeds 24 inches (610 mm). Entries and exits shall consist of one or a combination of the following: steps, stairs, ladders, treads, ramps, beach entries, underwater seats, benches, swimouts and other *approved* designs. The means of entry and exit shall be located on the shallow side of the first slope change.

809.3 Secondary entries and exits. Where water depth in the deep area of a pool exceeds 5 feet (1524 mm), a means of entry and exit as indicated in Section 809.2 shall be provided in the deep area of the pool.

> **Exception:** Where the required placement of a means of exit from the deep end of a pool would present a potential hazard, handholds shall be provided as an alternative for the means of exit.

809.4 Over 30 feet in width. Pools over 30 feet (9144 mm) in width at the deep area shall have an entry and exit on both sides of the deep area of the pool.

809.5 Pool stairs. The design and construction of stairs into the shallow end and recessed pool stairs shall conform to Sections 809.5.1 through 809.5.3.

809.5.1 Tread dimension and area. Treads shall have a minimum unobstructed horizontal depth of 10 inches (254 mm) and a minimum unobstructed walking surface area of 240 square inches (0.15 m²).

809.5.2 Riser heights. Risers, other than the top and bottom riser, shall have a uniform height of not greater than 12 inches (305 mm). The top riser height shall be any dimension not exceeding 12 inches (305 mm) for the width of the walking surface. The bottom riser height shall be any dimension not exceeding 12 inches (305 mm). The top and bottom riser heights shall not be required to be equal to each other or equal to the uniform riser height. Riser heights shall be measured at the horizontal centerline of the walking surface area.

809.5.3 Additional steps. In design water depths exceeding 48 inches (1219 mm), additional steps shall not be required.

809.6 Beach and sloping entries. The slope of beach and sloping entries used as a pool entrance shall not exceed 1 unit vertical in 7 units horizontal (14-percent slope).

809.7 Steps and sloping entries. Where steps and benches are used in conjunction with sloping entries, the vertical riser distance shall not exceed 12 inches (305 mm). For steps used in conjunction with sloping entries, the requirements of Section 809.6 shall apply.

809.8 Architectural features. Surfaces of architectural features shall not be required to comply with the 1 unit vertical in 7 units horizontal (14-percent slope) slope limitation.

809.9 Maximum depth. The horizontal surface of underwater seats, benches and swimouts shall be not greater than 20 inches (508 mm) below the *design waterline*.

SECTION 810—CIRCULATION SYSTEMS

810.1 Turnover rate. The circulation system equipment shall be sized to provide a turnover of the pool water not less than once every 12 hours. The system shall be designed to provide the required turnover rate based on the manufacturer's specified maximum flow rate of the filter, with a clean media condition of the filter.

810.2 Strainer required. Pressure filter systems shall be provided with a strainer located between the pool and the circulation pump.

SECTION 811—SAFETY FEATURES

811.1 Rope and float. In pools where the point of first slope break occurs, a rope and float assembly shall be installed across the width of the pool. The rope assembly shall be located not less than 1 foot (305 mm) and not greater than 2 feet (610 mm) towards the shallow side of the slope break. Rope anchoring devices shall be permanently attached to the pool wall, coping or deck. Rope ends shall attach to the rope anchor devices so that the rope ends can be disconnected from the rope anchor device.

PERMANENT RESIDENTIAL SPAS AND PERMANENT RESIDENTIAL EXERCISE SPAS

SECTION 901—GENERAL

901.1 Scope. This chapter shall govern the design, installation, construction and repair of permanently installed *residential* spas and exercise spas intended for *residential* use.

901.2 General. Permanent *residential* spas and permanent *residential* exercise spas shall comply with Chapter 5 except that Sections 504.1, 504.1.1 and 508.1 shall not apply. Such spas shall comply with the requirements of Chapter 3.

SECTION 902—SAFETY FEATURES

902.1 Instructions and safety signage. Instructions and safety signage shall comply with the manufacturer's recommendations and the requirements of the local jurisdiction.

PORTABLE RESIDENTIAL SPAS AND PORTABLE RESIDENTIAL EXERCISE SPAS

User notes:

About this chapter: *Chapter 10 regulates portable residential spas and exercise spas by requiring compliance with product standards.*

SECTION 1001—GENERAL

1001.1 Scope. This chapter shall govern the installation, *alteration* and repair of portable *residential* spas and portable exercise spas intended for *residential* use.

1001.2 General. In addition to the requirements of this chapter, portable *residential* spas and portable *residential* exercise spas shall comply with the requirements of Chapter 3.

1001.3 Listing. Equipment and appliances shall be *listed* and *labeled*, and installed as required by the terms of their approval, in accordance with the conditions of the listing, the manufacturer's instructions and this code. Manufacturer's instructions shall be available on the job site at the time of inspection.

1001.4 Certification. Factory-built portable spas and portable exercise spas installed in *residential* applications shall be *listed* and *labeled* in compliance with UL 1563 or CSA C22.2 No. 218.1.

1001.5 Installation. Spa equipment shall be supported to prevent damage from misalignment and settling in accordance with the manufacturer's instructions.

1001.6 Access. Electrical components that require placement or servicing shall be located with access.

1001.7 Instructions and safety signage. Instructions and safety signage shall comply with UL 1563 or CSA C22.2 No. 218.1, the manufacturer's recommendations, and the requirements of the local jurisdiction.

REFERENCED STANDARDS

User notes:

About this chapter: *This code contains numerous references to standards promulgated by other organizations that are used to provide requirements for materials and methods of construction. Chapter 11 contains a comprehensive list of all standards that are referenced in this code. These standards, in essence, are part of this code to the extent of the reference to the standard.*

This chapter lists the standards that are referenced in various sections of this document. The standards are listed herein by the promulgating agency of the standard, the standard identification, the effective date and title, and the section or sections of this document that reference the standard. The application of the referenced standards shall be as specified in Section 102.7.

ACI   *American Concrete Institute, 38800 Country Club Drive, Farmington Hills, MI 48331-3439*

ACI 318—19: Building Code Requirements for Structural Concrete
> Table 307.2.2

AHRI   *Air Conditioning, Heating & Refrigeration Institute, 2311 Wilson Boulevard, Suite 400, Arlington, VA 22201*

400 (I-P)—2015: Performance Rating of Liquid to Liquid Heat Exchangers
> Table 317.2(1)

1160 (I-P)—2022: Performance Rating of Heat Pump Pool Heaters—(with Addendum 1)
> Table 317.2(1)

ANSI   *American National Standards Institute, 25 West 43rd Street, 4th Floor, New York, NY 10036*

A108/A118/A136.1—2019: Specifications for Installation of Ceramic Tile
> Table 307.2.2

A326.3—18: Test Method for Measuring Dynamic Coefficient of Friction for Hard Surface Floor Materials
> 306.2

Z21.56a:19/CSA 4.719: Gas Fired Pool Heaters
> Table 317.2(1)

APSP   *Pool & Hot Tub Alliance (formerly The Association of Pool & Spa Professionals), 2111 Eisenhower Avenue, Suite 500, Alexandria, VA 22314*

ANSI/APSP/ICC 4—2012: American National Standard for Aboveground/Onground Residential Swimming Pools—Includes Addenda A Approved April 4, 2013
> 702.2.1

ANSI/APSP/ICC 14—2019: American National Standard for Portable Electric Spa Energy Efficiency
> 303.2

ANSI/APSP/ICC 16—2017: American National Standard for Suction Outlet Fitting Assemblies (SOFA) for Use in Pools, Spas, and Hot Tubs
> 202, 312.4.1, 312.4.4, 505.2.1, 612.4.2

ANSI/APSP/ICC/NPC 12—2016: American National Standard for the Plastering of Swimming Pools and Spas
> 307.2.5

ANSI/PHTA/ICC 7—2020: American National Standard for Suction Entrapment Avoidance in Swimming Pools, Wading Pools, Spas, Hot Tubs, and Catch Basins
> 311.1

ANSI/PHTA/ICC 15—2021: American National Standard for Residential Swimming Pool and Spa Energy Efficiency
> 303.3

ASCE/SEI   *American Society of Civil Engineers Structural Engineering Institute, 1801 Alexander Bell Drive, Reston, VA 20191-4400*

ASCE/SEI 24—14: Flood Resistant Design & Construction
> 304.3

ASHRAE
ASHRAE, 180 Technology Parkway, Peachtree Corners, GA 30092

62.1—2022: Ventilation for Acceptable Air Quality
326.1

ASME
American Society of Mechanical Engineers, Two Park Avenue, New York, NY 10016-5990

A112.1.2—2012 (R2022): Air Gaps in Plumbing Systems (For Plumbing Fixtures and Water-connected Receptors)
319.2

B16.15—2023: Cast Alloy Threaded Fittings: Classes 125 and 250
Table 312.4.1

ASTM
ASTM International, 100 Barr Harbor, P.O. Box C700, West Conshohocken, PA 19428-2959

A182/A182M—21: Standard Specification for Forged or Rolled Alloy and Stainless Steel Pipe Flanges, Forged Fittings, and Valves and Parts for High-temperature Service
Table 312.4.1

A240/A240M—20a: Standard Specification for Chromium and Chromium-nickel Stainless Steel Plate, Sheet and Strip for Pressure Vessels and for General Applications
Table 307.2.2

A312/A312M—21: Standard Specification for Seamless, Welded, and Heavily Cold Worked Austenitic Stainless Steel Pipes
Table 312.4

A403/A403M—20: Standard Specification for Wrought Austenitic Stainless Steel Piping Fittings
Table 312.4.1

B88—20: Standard Specification for Seamless Copper Water Tube
Table 312.4

B447—12a(2021): Specification for Welded Copper Tube
Table 312.4

D1527—99(2005): Specification for Acrylonitrile Butadiene Styrene (ABS) Plastic Pipe, Schedules 40 and 80
Table 312.4, Table 312.4.1

D1593—19: Standard Specification for Nonrigid Vinyl Chloride Plastic Film and Sheeting
Table 307.2.2

D1785—21a: Specification for Poly (Vinyl Chloride) (PVC) Plastic Pipe, Schedules 40, 80 and 120
Table 312.4

D2241—21: Standard Specification for Poly (Vinyl Chloride) (PVC) Pressure-rated Pipe (SDR Series)
Table 312.4

D2464—15: Standard Specification for Threaded Poly (Vinyl Chloride) (PVC) Plastic Pipe Fittings, Schedule 80
Table 312.4.1

D2466—21: Standard Specification for Poly (Vinyl Chloride) (PVC) Plastic Pipe Fittings, Schedule 40
Table 312.4.1

D2467—20: Standard Specification for Poly (Vinyl Chloride) (PVC) Plastic Pipe Fittings, Schedule 80
Table 312.4.1

D2672—20e1: Standard Specification for Joints for IPS PVC Pipe Using Solvent Cement
Table 312.4

D2846/D2846M—19a: Standard Specification for Chlorinated Poly (Vinyl Chloride) (CPVC) Plastic Hot- and Cold-Water Distribution Systems
Table 312.4, Table 312.4.1

D3787—16(2020): Standard Test Method for Bursting Strength of Textiles—Constant-Rate-of-Traverse (CRT) Ball Burst Test
305.2.4.1

D5034—09(2017): Standard Test Method for Breaking Strength and Elongation of Textile Fabrics (Grab Test)
305.2.4.1

F437—21: Standard Specification for Threaded Chlorinated Poly (Vinyl Chloride) (CPVC) Plastic Pipe Fittings, Schedule 80
Table 312.4.1

F438—2017: Standard Specification for Socket-type Chlorinated Poly (Vinyl Chloride) (CPVC) Plastic Pipe Fittings, Schedule 40
Table 312.4.1

F439—19: Standard Specification for Chlorinated Poly (Vinyl Chloride) (CPVC) Plastic Pipe Fittings, Schedule 80
Table 312.4.1

F1346—1991(2018): Standard Performance Specification for Safety Covers and Labeling Requirements for All Covers for Swimming Pools, Spas and Hot Tubs
305.1, 305.4

F2286—16: Standard Design and Performance Specification for Removable Mesh Fencing for Swimming Pools, Hot Tubs, and Spas
305.2.5

CPSC   *Consumer Product Safety Commission, 4330 East-West Highway, Bethesda, MD 20814*

16 CFR Part 1207 (2004): Safety Standard for Swimming Pool Slides
406.10

CSA   *CSA Group, 8501 East Pleasant Valley Road, Cleveland, OH 44131-5516*

B137.2—23: Polyvinylchloride (PVC) Injection-moulded Gasketed Fittings for Pressure Application
Table 312.4, Table 312.4.1

B137.3—23: Rigid Polyvinylchloride (PVC) Pipe and Fittings for Pressure Applications
Table 312.4, Table 312.4.1

B137.6—23: Chlorinated Polyvinylchloride (CPVC) Pipe, Tubing, and Fittings for Hot- and Cold-water Distribution Systems
Table 312.4, Table 312.4.1

C22.2 No. 108—14(R2019): Liquid Pumps
314.8

C22.2 No. 218.1—13 (2017): Spas, Hot Tubs and Associated Equipment
302.3, 310.1, 311.1, 314.8, Table 317.2(1), 318.2, 1001.4, 1001.7

C22.2 No. 236—15: Heating and Cooling Equipment
Table 317.2(1)

Z21.56a—19/CSA 4.7—19: Gas Fired Pool Heaters
Table 317.2(1)

IAPMO   *IAPMO, 4755 E. Philadelphia Street, Ontario, CA 91761-USA*

IAPMO Z124.7—2013: Prefabricated Plastic Spa Shells
Table 307.2.2

ICC   *International Code Council, Inc., 200 Massachusetts Avenue, NW, Suite 250, Washington, DC 20001*

IBC—24: International Building Code®
201.3, 304.2, 306.1, 307.2, 410.1

ICC 901/SRCC 100—2020: Solar Thermal Collector Standard
317.6.2

ICC 902/APSP/SRCC 400—2020: Solar Pool and Spa Heating System Standard
Table 317.2(2)

ICC A117.1— 2017: Standard for Accessible and Usable Buildings and Facilities
307.1.5

IECC—24: International Energy Conservation Code®
201.3, 317.4

IFC—24: International Fire Code®
201.3,

IFGC—24: International Fuel Gas Code®
201.3, , 317.4,

IMC—24: International Mechanical Code®
201.3, , 317.4,

IPC—24: International Plumbing Code®
201.3, 302.2, 302.5, 302.6, 306.8, 319.2, 410.1

IRC—24: International Residential Code®
102.7.1, 201.3, 302.1, 302.5, 302.6, 304.2, 306.1, 306.3, 306.8, 307.2, 317.4, 319.2, 322.2.1, 322.4, 703.1, 802.1, 802.2

NEMA   *National Electrical Manufacturers Association, 1300 North 17th Street Suite 900, Rosslyn, VA 22209*

NEMA Z535—2017: ANSI/NEMA Color Chart
409.3

NFPA   *National Fire Protection Association, 1 Batterymarch Park, Quincy, MA 02169-7471*

NFPA 70—2023: National Electrical Code
302.1, 317.4, 322.2.1, 322.4

NSF   *NSF International, 789 N. Dixboro Road P.O. Box 130140, Ann Arbor, MI 48105*

NSF 14—2020: Equipment and Chemicals for Swimming Pools, Spas, Hot Tubs and Other Recreational Water Facilities
302.3, 312.4

NSF 50—2020: Equipment for Swimming Pools, Spas, Hot Tubs, and Other Recreational Water Facilities
302.3, 310.2, 508.1, 612.5.4

PHTA   *Pool and Hot Tub Alliance, 2111 Eisenhower Avenue, Suite 500, Alexandria, VA 22314*

ANSI/PHTA/ICC 10—2021: American National Standard for Elevated Pools and Spas
309.1

SA   *Standards Australia, Level 10, The Exchange Centre, 20 Bridge Street, Sydney, Australia*

AS 4586—2013: Slip resistance classification of new pedestrian surface materials
306.2

UL   *UL LLC, 333 Pfingsten Road, Northbrook, IL 60062*

372—2007: Automatic Electrical Controls for Household and Similar Use—Part 2: Particular Requirements for Burner Ignition Systems and Components—with revisions through June 2012
506.2.1, 506.2.2

873—2007: Temperature-indicating and Regulating Equipment—with revisions through February 2015
506.2.1, 506.2.2

1004-1—12: Rotating Electrical Machines General Requirements—with revisions through November 2020
314.8

1081—2016: Swimming Pool Pumps, Filters and Chlorinators—with revisions through July 2020
314.8

1261-- 2001: Electric Water Heaters for Pools and Tubs—with Revisions through September 2017
Table 317.2(1)

1563—2009: Electric Spas Equipment Assemblies and Associated Equipment—with revisions through September 2020
302.3, 310.1, 311.1, 314.8, Table 317.2(1), 318.2, 1001.4, 1001.7

1995—2015: Heating and Cooling Equipment—with revisions through August 2018
Table 317.2(1)

2017—2008: General-purpose Signaling Devices and Systems—with revisions through December 2016
305.4

60335-2-1000-17: Household and Similar Electrical Appliances: Particular Requirements for Electrically Powered Pool Lifts
307.1.5

UL/CSA 60335-2-40—2022: Standard for Safety for Household and Similar Electrical Appliances—Safety—Part 2-40: Particular Requirements for Electrical Heat Pumps, Air-Conditioners and Dehumidifiers
Table 317.2(1)

The provisions contained in this appendix are not mandatory unless specifically referenced in the adopting ordinance.

User notes:

About this appendix: *Appendix C provides criteria for Board of Appeals members. Also provided are procedures by which the Board of Appeals should conduct its business.*

Code development reminder: *Code change proposals to this appendix will be considered by the Administrative Code Development Committee during the 2025 (Group B) Code Development Cycle.*

SECTION A101—GENERAL

[A] A101.1 Scope. A board of appeals shall be established within the jurisdiction for the purpose of hearing applications for modification of the requirements of this code pursuant to the provisions of Section 112. The board shall be established and operated in accordance with this section and shall be authorized to hear evidence from appellants and the *code official* pertaining to the application and intent of this code for the purpose of issuing orders pursuant to these provisions.

[A] A101.2 Application for appeal. Any person shall have the right to appeal a decision of the *code official* to the board. An application for appeal shall be based on a claim that the intent of this code or the rules legally adopted hereunder have been incorrectly interpreted, the provisions of this code do not fully apply or an equally good or better form of construction is proposed. The application shall be filed on a form obtained from the *code official* within 20 days after the notice was served.

[A] A101.2.1 Limitation of authority. The board shall not have authority to waive requirements of this code or interpret the administration of this code.

[A] A101.2.2 Stays of enforcement. Appeals of notice and orders, other than Imminent Danger notices, shall stay the enforcement of the notice and order until the appeal is heard by the board.

[A] A101.3 Membership of board. The board shall consist of five voting members appointed by the chief appointing authority of the jurisdiction. Each new member shall serve for **[NUMBER OF YEARS]** years or until a successor has been appointed. The board member's terms shall be staggered at intervals, so as to provide continuity. The *code official* shall be an ex officio member of said board but shall not vote on any matter before the board.

[A] A101.3.1 Qualifications. The board shall consist of five individuals, who are qualified by experience and training to pass on matters pertaining to building construction and are not employees of the jurisdiction.

[A] A101.3.2 Alternate members. The chief appointing authority is authorized to appoint two alternate members who shall be called by the board chairperson to hear appeals during the absence or disqualification of a member. Alternate members shall possess the qualifications required for board membership, and shall be appointed for the same term or until a successor has been appointed.

[A] A101.3.3 Vacancies. Vacancies shall be filled for an unexpired term in the same manner in which original appointments are required to be made.

[A] A101.3.4 Chairperson. The board shall annually select one of its members to serve as chairperson.

[A] A101.3.5 Secretary. The chief appointing authority shall designate a qualified clerk to serve as secretary to the board. The secretary shall file a detailed record of all proceedings that shall set forth the reasons for the board's decision, the vote of each member, the absence of a member and any failure of a member to vote.

[A] A101.3.6 Conflict of interest. A member with any personal, professional or financial interest in a matter before the board shall declare such interest and refrain from participating in discussions, deliberations and voting on such matters.

[A] A101.3.7 Compensation of members. Compensation of members shall be determined by law.

[A] A101.3.8 Removal from the board. A member shall be removed from the board prior to the end of their terms only for cause. Any member with continued absence from regular meeting of the board may be removed at the discretion of the chief appointing authority.

[A] A101.4 Rules and procedures. The board shall establish policies and procedures necessary to carry out its duties consistent with the provisions of this code and applicable state law. The procedures shall not require compliance with strict rules of evidence, but shall mandate that only relevant information be presented.

[A] A101.5 Notice of meeting. The board shall meet, upon notice from the chairperson, within 10 days of the filing of an appeal or at stated periodic intervals.

[A] A101.5.1 Open hearing. All hearings before the board shall be open to the public. The appellant, the appellant's representative, the *code official* and any person whose interests are affected shall be given an opportunity to be heard.

[A] A101.5.2 Quorum. Three members of the board shall constitute a quorum.

[A] A101.5.3 Postponed hearing. When five members are not present to hear an appeal, either the appellant or the appellant's representative shall have the right to request a postponement of the hearing.

[A] A101.6 Legal counsel. The jurisdiction shall furnish legal counsel to the board to provide members with general legal advice concerning matters before them for consideration. Members shall be represented by legal counsel at the jurisdiction's expense in all matters arising from service within the scope of their duties.

[A] A101.7 Board decision. The board shall only modify or reverse the decision of the *code official* by a concurring vote of three or more members.

[A] A101.7.1 Resolution. The decision of the board shall be by resolution. Every decision shall be promptly filed in writing in the office of the *code official* within 3 days and shall be open to the public for inspection. A certified copy shall be furnished to the appellant or the appellant's representative and to the *code official*.

[A] A101.7.2 Administration. The *code official* shall take immediate action in accordance with the decision of the board.

[A] A101.8 Court review. Any person, whether or not a previous party of the appeal, shall have the right to apply to the appropriate court for a writ of certiorari to correct errors of law. Application for review shall be made in the manner and time required by law following the filing of the decision in the office of the chief administrative officer.

INDEX

codes.iccsafe.org

Get the Latest PMG Codes on Your Digital Device

Subscribe to the ICC Digital Codes PMG Collection

The ICC Digital Codes Premium PMG Collection offers a range of the latest Plumbing, Mechanical, Fuel Gas, Swimming Pool & Spa, (PMG) codes and reference standards. It also includes related support materials such as commentaries, user guides, study companions, significant code changes and revision history.

PMG COLLECTION

Subscribe to the PMG Collection and unlock exclusive features

- ✓ Classify notes, files and videos into relevant code sections
- ✓ Share your access and content simultaneously
- ✓ View past committee code interpretations
- ✓ View full change history and public comments
- ✓ View real time code change proposals
- ✓ View updates and significant codes changes to I-Codes

LEARN MORE AT CODES.ICCSAFE.ORG

ES ICC EVALUATION SERVICE®

ICC-ES PMG is the leading provider of product certifications in plumbing, mechanical and fuel gas.

✓ **Excellent customer service**

✓ **Highest acceptability by code officials**

✓ **The best value**

We certify:
- Kitchen Cabinets
- Faucets
- Showerheads
- Bathtubs
- Sink products
- Toilets & Urinals
- Ceramic Fixtures
- Resins and Linear Drain products

and many more!

Benefits of having an ICC-ES PMG Listing:

- ICC-ES PMG offers a **lower cost** for certification than competitors

- Expedited certification for all client listings

- ICC-ES PMG **does not conduct warehouse inspections**

- ICC-ES PMG **does not charge for additional company listings**

- ICC-ES will accept test reports from other entities

- **No fee for EPA WaterSense listings and lead law listings**

- **No separate file for NSF 61 listings**

- A2LA, ANSI, JAS-ANZ, SCC and EMA accreditation

www.icc-es.org/pmg
800-423-6587 x7643

23-22253

CREDENTIALING

ICC Credentialing provides nationally recognized credentials that demonstrate a confirmed commitment to protect public health, safety, and welfare. Raise the professionalism of your department and further your career by pursuing an ICC Certification.

ICC Certifications Offer

- Nationwide recognition
- Increased earning potential
- Career advancement
- Superior knowledge
- Validation of your expertise
- Personal and professional satisfaction

Exams are developed and maintained to the highest standards, which includes continuous peer review by national committees of experienced, practicing professionals.

High Demand Exams (exam ID)

- Commercial Plumbing Inspector (P2)
- Plumbing Plans Examiner (P3)
- Commercial Mechanical Inspector (M2)
- Mechanical Plans Examiner (M3)
- Residential Building Inspector (B1)
- Commercial Building Inspector (B2)
- Building Plans Examiner (B3)

ICC ONLINE EXAMS

Proctored Remote Online Testing Option (PRONTO) provides a convenient testing experience that is accessible 24 hours a day, 7 days a week, 365 days a year.

Required hardware/software is minimal – you will need a webcam and microphone, as well as a reasonably recent operating system.

Checkout all that ICC Credentialing has to offer at
iccsafe.org/certification

23-22499

INTERNATIONAL CODE COUNCIL®

Changes to the Code Development Process

The International Code Council is revising its code development process to improve the quality of code content by fostering a more in-depth vetting of code change proposals. Changes will take effect in 2024–2026 for the development of the 2027 I-Codes and will move the development process to an integrated and continuous three-year cycle.

In the new timeline:

- **Year 1** will include two Committee Action Hearings for Group A Codes·

- **Year 2** will include two Committee Action Hearings for Group B Codes

- **Year 3** will be the joint Public Comment Hearings and Online Governmental Consensus Vote for both Group A and B Codes

Learn more about the new process and download the new timeline here.

23-22529